TRAGEDY AT HONDA

Tragedy at Honda

By Charles A. Lockwood

and

Hans Christian Adamson

Tragedy at Honda by Charles A. Lockwood and Hans Christian Adamson. Published 2019 by Binnacle Books.

FIRST PRINTING, 2018.

ISBN: 9781691326259.

CONTENTS

1 – HONDA AND THE DEVIL'S JAW

HONDA.

BLACK, BLEAK, AND HOSTILE; the scene of countless tragedies of
the sea; Honda mesa lifts its forbidding cliffs of primordial volcanic rock steeply out of Magellan's misnamed Pacific Ocean. It
rears its ugly head about 15 miles northwest of the point where
California's coastal sea lanes bend sharply to the east to enter
Santa Barbara Channel. Extending seaward from the base of the
craggy bluff which at its highest is as tall as an eight-story building; a sweeping semicircle of serrated rocks, needle-sharp
pinnacles, and razor-honed reefs stands boldly above the water
or lies hidden below its surface. Without stretching the imagination too far, the scene resembles the fang-studded jaw of a
grotesque prehistoric monster built on a gigantic scale.

Honda.

A weird and depressing place. Brooding silence and the life-
lessness of a morgue spread like a pall over the desolate spot
when the wind is low and the sea is quiet. Not even the calls of
gulls, or of other marine birds, are heard. The only sound, the
only movement, is the endless cycle of the surf; the eternal slith-
ering of waves that rise and strike against the bluff, the rocks,
and the pinnacles only to recede—hissing angrily; like broods of
frustrated serpents of the deep.

Honda.

Lair of evil jinns who serve the despotic ocean in its most ty-
rannical and destructive moods. The jinns of unpredictable gales
that maul, break, and rip obstacles in their paths. The jinns of
fog, gray and dense as dirty wool, that wipe out vision and muffle
sound. The jinns of wayward currents that carry seafaring men to
disaster. The jinns of high-crested seas that form regiments of
thundering breakers. When the jinns of Honda do the bidding of
the cruel sea, mighty forces are set into motion.

Far offshore, winds come howling out of unexpected corners.
Along the coast, impenetrable fog hides the bluff as well as its offshore reefs and rocks. In the shipping lanes, uncharted currents
with onshore sets begin to flow. And, at Honda, spume-crested
seas hurl themselves upon cliffs and rocks amid crashing roars
like those of a thousand cannon. Then Honda is in its element.

The trap is set: waiting, patient, and ready for its next victim. Be the offering large or small, steel or timber, it makes no difference. All ships are welcome morsels to its insatiable appetite.

Honda.

An ancient and enduring menace to ships and seafarers from the sixteenth century until this very day. The Spaniards, who plied their tall-masted, high-pooped galleons between Mexico and the Philippines, had a name for Honda: they called it La Guijada del Diablo, 'The Devil's Jaw.'

Today's men of the sea know it as the Graveyard of Ships. Generations of sailors have spoken the word Honda with profound respect; thousands of them with soul-gripping fear; hundreds of them in mortal agony.

And yet, the name Honda does not appear on any mariner's chart or on any surveyor's map. As a means of geographical identification it is written mainly in the minds of the men of the sea.

And, even then, it is not always recorded in the same manner. To some, it is known as Point Honda. To others as Honda Head. To still others as La Honda or Honda Mesa; often as not, just Honda. Actually, Honda is part of a landmass that forms a bulging bluff listed on charts and maps as Point Pedernales. It is located less than 3 miles north of Point Arguello. There the flashing beam and the hoarse fog signal of Point Arguello Light and the electronic bearings broadcast by the Arlight Naval Radio Station do their share toward making the coastal lanes safe for seafarers who, in weather fair or foul, shape their course to make the sharp turn to port from the open ocean into Santa Barbara Channel.

At low tide, the rock-strewn beaches before Honda reveal the gruesome tomb markers of bleached bones and rusty remnants of ships that somehow made the critical turn too soon, only to be caught in the grip and chewed to bits in the Devil's Jaw.

Since 1542, when Juan Cabrillo was the first European to sail along the California coast, Honda has taken a steady and shocking toll of coastal shipping. It was thus in the days of sail. And it has continued to be thus in this century of expanding nuclear propulsion.

No one knows how Honda got its name. But it is believed that it was first bestowed upon the barren and desert-like mesa which rises gradually from the cliffs that face the sea and runs eastward to the foothills of the Santa Ynez Mountains. Backdrop of the Honda scene is a huge lump of a rock that rises to a height of 2100 feet and bears the inappropriate name of Mount Tranquillon. The northern boundary of Honda mesa is rimmed by a gully which early Spanish ranchers, because of its depth, called 'Canada Hondo.' Literally translated, this means 'Deep Trail.'

In time, so speculation runs, this was shortened to Honda and made to include not only the mesa but also the ocean's diabolical trap of explosive violence and sudden death. A name to be feared by captains of vessels large and small, sail and steam, merchant ships and men-o'-war...

Honda!

U.S.S. WOODBURY AT THE POINT HONDA SHIPWRECK SITE

2 - So Long, San Francisco!

OVER THE STEEPLES AND THE TOWERS that grace the hills of the City by the Golden Gate, a favorite haven and host to seafaring men of all nations since the seventeen hundreds, the dawn of September 8, 1923, broke clear and warm because September is virtually the sunny season of San Francisco's unorthodox climate. The accustomed fog, which plagued seagoing operations and ferry traffic during the hours of darkness, had lifted. The eastern sky was aflame with red as the sun tinged the edges of a few wind-blown black clouds. Soon it rose majestically over Mount Diablo to open and adorn the day as well as to drive the shadows of night from the cluttered waterfronts and streets of San Francisco and her sister cities that rim the Bay.

Probably few of those awake to witness Nature's thrilling color display on the birth of this new day gave even a passing thought to that age-old sailorman's warning: "Red sky in the morning, sailors take warning," an omen of death and disaster at sea.

Hoary maxims of that kind belonged to the age of sail. In this day of steam and of steel-hulled ships, with thousands of horsepower awaiting a touch of the throttle, dangers from storm, wind, and wave have come to mean little to modern seamen. Or so they seem to think. The thoughts of watch-standing naval personnel on the decks and bridges of the large and small, shark-gray men-o'-war which crowded the docks and mid-stream anchorage were probably concerned with the fact that another Fleet Week in hospitable, fascinating San Francisco had passed into history.

Yes: Fleet Week, with its naval review, its fanfare, and its dances; with its parades, its boat races and baseball games; its thundering salutes to visiting dignitaries, snappy side-boys, and the shrill beeps of the boson's pipe, was over. The mighty battleships of the United States Fleet, after a summer of Fleet Exercises in Puget Sound and northern areas, were even now, with attendant destroyer squadrons and auxiliary vessels, getting up steam for return to their regular bases at San Pedro and San Diego. The destroyer Young was a trim, smart-looking member of Captain Robert Morris' top-notch outfit, Destroyer Division 33.

She was nested with her division mates at Pier 15 in San Francisco.

On her deck, two men stood talking at the gangway. One, the Officer of the Watch, was a young ensign. Complete with binoculars slung around his neck and, in his hand, a cup of coffee strong enough to float a depth charge, he listened with evident interest to his Quartermaster of the Watch, a veteran of many cruises in "four-stackers" or "tin-cans," as they were more familiarly known in Navy slang. Idly passing the last few minutes before the time came to call all hands and get the ship ready for sea, the duo had been speculating upon the portent of such a blood-red sunrise. The Quartermaster had seemed worried.

"Yes, sir," the older man was saying, "I know it's a mighty red sky, but that's not what's bothering me.

I've seen lots of threatening sunrises that didn't bring on stormy weather. I guess it's the fact that our new Chief Commissary Steward—a good man with the menus, who kept the cooks on their toes and fed us swell chow—has been absent over liberty for two days. See? Unless he shows up in the next half hour, he won't make this trip to San Diego with the Young. See, sir?"

"Yes, I see!" replied the OOD as he suppressed a grin. "Lieutenant Donaldson, the Commissary Officer, also has been worried about that. He's checked with the contractors ashore and inventoried the store-rooms. The Chief's accounts are straight, the provision lockers well stocked—and we won't starve."

"Well, sir, the truth is, according to scuttlebutt, that he's been jittery about making this run down the coast. He is supposed to have said he had a hunch something's going to happen—something bad. See? Guess he's got some of the boys believing him. Me! I don't quite know what to think. But ..."

"What! An old hand like you superstitious, Quartermaster?" scoffed the Ensign, as he smiled incredulously.

"No, sir and I actually hadn't thought about it until yesterday afternoon. But on the First Dog Watch, I saw three or four rats trying to get ashore from our ship over the mooring lines and gangway. See? One of the black gang kicked two of them overboard. And, sir, you know that rats quitting the ship is a jolt to any sailor."

From the distant Oakland waterfront, interrupting their talk, came the muted sound of sirens and whistles being tested. It was echoed from the piers of San Francisco as other ships of Destroyer Squadron 11 tested their sound signal equipment preparatory to getting underway. As per routine, the four-stackers—the "Cavalry of the Sea," as they liked to consider themselves—were to precede the Fleet out the channel.

Theoretically, their purpose was to liquidate lurking enemy submarines. In the rush of getting underway, the conversation with the Quartermaster was forgotten. However, the Ensign OOD recalled later that, as lines were being singled up, he saw no rats deserting the ship. The Chief Commissary Steward, however, was still missing as the Young backed away from the "nest."

DesDiv 33 backed out of Pier 15 and formed up into Division, with the flagship S. P. Lee leading, followed by the Young, the Woodbury, and the Nicholas. At the same time, DesDiv 31, commanded by Commander William S. Pye, was snaking its way out of Oakland Estuary from its berth at the Municipal Pier. The Farragut, as Division Flag, led the way with the Somers, Fuller, J. F. Burnes, Percival, and Chauncey tailed in her wake.

Upon signal to clear the harbor, DesDiv 32, commanded by Commander Walter G. Roper and nested at Pier 36 on the Embarcadero, backed out of its berth. One by one, its DDs formed up with the flagship Kennedy leading. The Thompson, the Paul Hamilton, and the Stoddert followed in that order. The Delphy, Flagship of Captain Edward H. Watson, Commander Destroyer Squadron 11—composed of DesDivs 31, 32, and 33—after the preliminary shrill "Whripp, whripp, whripp" of her siren and the deep-throated "Whoot" of her whistle, got underway from her anchorage in mid-channel, with signal flags flying from both yardarms.

Meanwhile Destroyer Squadron 12—Captain James H. Tomb—scheduled to follow DesRon 11 to sea, had been clearing the throats of its whistles and sirens. For late sleepers in San Francisco, it was a noisy morning. There were four absentees from the destroyers of Captain Watson's command on that beautiful September morning. The William Jones, of DesDiv 33, skippered by Lieutenant Commander B. B. Taylor, had suffered an engineering casualty which prevented her from working up to the higher speeds. She therefore had got underway shortly after midnight, following, through heavy fog, the movements of the destroyer tender Melville—flagship of DesRons—which also proceeded to sea at that hour.

Rear Admiral S. E. W. Kittelle, Commander Destroyer Squadrons, Battle Fleet, who flew his two-starred flag in the Melville, departed at this ghostly hour because of his Flagship's limited speed and in order to reach the Destroyer Base in San Diego not too long after his more fleet-footed fighting ships arrived in their home port. The Farquhar of DesDiv 32, commanded by Lieutenant Commander Jeff Davis Smith, was another cripple and got underway on the stroke of midnight to trail the fog whistle of the Melville out of the Bay.

The Reno, also of DesDiv 32, backed away from her mates immediately after the Farquhar and headed seaward, amid the mournful hooting of fog signals, but not because of engineering difficulties. On the contrary, her genial, good-looking skipper, Lieutenant Commander J. R. "Dick" Barry, a veteran destroyer man, had requested and received orders to take advantage of this operation to make the annual smoke prevention and full-speed run required of all destroyers. His zeal and foresight was commendable and deserved success. However, Fate decreed otherwise, for the test was never completed. Still, from the human angle, the unforeseen outcome reflected great credit upon the Reno and her people.

The final absentee was from DesDiv 33, the Zeilin, commanded by Lieutenant Commander H. G. Shonerd. She was, at this time, in dry dock at Seattle because of a near-fatal collision in Puget Sound.

This destroyer was one of the honor escorts of the Navy Transport Henderson which, in a dense fog, at 0756 on July 27, had rammed the escorting destroyer on the port side amidships.

So serious was the flooding which resulted that all hands abandoned the Zeilin at 0815 in anticipation of capsizing.

Only a few days before, while hurrying to Seattle to arrive before the Henderson, the heavy cruiser Seattle (Flagship of the United States Fleet, with the Commander-in-Chief, Admiral Robert E. Coontz, aboard) had gone aground in one of those pea soup fogs for which that region is famous.

The Henderson, the unwitting cause of these calamities, was returning to Seattle after taking the President of the United States, Warren G. Harding, on a round trip to Alaska for an inspection of that land of fog, fish, snow, ice, huskies, gold, and potential oil.

Six days later, in San Francisco on August 2, President Harding died. In the over-all picture, it had not been a happy summer for the United States Fleet. But, in this first week of September, better times appeared to lie ahead. Naval appropriations, made by the Harding Administration, for the fiscal year 1923-24 had been more liberal. This would permit bringing the short-handed crews of naval vessels up to full, or nearly full, complements.

It also permitted increased fuel allowances for all ships, thus providing for more realistic and more frequent battle exercises and training. This news was especially welcome to the faster ships of the Navy—carriers, cruisers, and destroyers. Especially destroyers, the very essence of whose almost suicidal torpedo attacks on enemy battle lines is speed—speed to get into torpedo firing range—speed to get out again before being battered to smoking hulks by the flaming guns of the enemy.

In the final phase of their summer cruise, the forthcoming run to San Diego, Admiral Kittelle thought he saw an excellent opportunity to use to good advantage some of this increased fuel allowance for a high-speed endurance test of his destroyers' engineering plants. Such a realistic trial would immediately show which ships were ready for high-speed battle operations and which ships needed improved maintenance and upkeep methods—or, possibly, new engineer officers or new skippers.

Destroyer men are, of necessity, very realistic in their approach to the solution of problems. Weak links do not strengthen a chain or a team, such as a destroyer squadron must be.

And so it was that, when Admiral Kittelle's operation order directing the return of Destroyer Squadrons, Battle Fleet, to San Diego was issued, it specified that all destroyers, with the exception of a few cripples, should conduct an endurance run to their home base. Captain E. H. Watson, ComDesRon 11, and Captain J. H. Tomb, ComDesRon 12, passed the appropriate orders down the line to their respective commands. As was customary before embarking upon an operation of importance, Captain Watson called for a conference of Commanding and Engineer Officers on the afternoon of September 7, the day before DesRon 11 was scheduled to sortie from San Francisco.

The Division Commanders, of course, likewise were summoned. Realizing that the tiny wardroom of the Delphy could not comfortably hold the 40-some officers who would attend, Captain Watson requested, and was granted, permission by Captain B. B. "Buzzing Benny" Wygant of the Melville, along whose port side Delphy was moored, to use the flagship's much larger wardroom.

Shortly before 1500, the Melville's starboard gangway was besieged by a flotilla of tiny destroyer gigs. With smart speed, they discharged their loads of sun-tanned, brine-washed, sharp-looking destroyer officers. The keenness of their glances and the roll of their walk marked them as men long accustomed to the vicissitudes of life, wind, and wave aboard their high-speed and versatile vessels.

Captain Watson—"Commodore" Watson, as Squadron and Division Commanders are traditionally called—receiving them in the wardroom, took his place at the head of the long table and called the conference to order on the stroke of 6 bells.

As he returned their greetings and looked down the lines of faces turned toward him, pride showed in his every feature and in the tone of his voice. While he was widely regarded as a battleship skipper, much of his career had been spent in smaller ships. These men were of the type that he knew: competent, resourceful, undaunted. During the 13 months that he had commanded DesRon 11, Watson had come to know their many fine qualities.

Present were his principal staff officers, Lieutenant Commander H. G. "Blinky" Donald, Squadron Engineer Officer; Lieutenant Commander H. K. "Chink" Lewis, Squadron Gunnery Officer; and Lieutenant Laurence "Jasper" Wild, Squadron Communications Officer. All were experienced destroyer sailors and excellent in their specialties.

"Gentlemen," said the Squadron Commander quietly, "you have all received Commander Destroyer Squadrons, Battle Fleet, Operation Order 39 dash 23 of August 31. You know the times set for getting underway and the order in which we will sortie. Divisions will proceed out of the harbor independently. DesRon 12 will follow us out and proceed independently to San Diego. It will keep well to westward of us, so that both squadrons will have room for maneuvering.

"I expect to form the Squadron off San Francisco Lightship at 0830 on a southerly course, in line of divisions. We will have a few tactical maneuvers and short-range battle practice training."

He looked around the room and smiled as he paused. Then he continued; "The only part of the operation order which may prove a bit difficult is the matter of the standard speed specified. Commander Donald will have a few comments to make on that part. This has been a tough summer and I know you have been concerned, as I have, over the number of engineering casualties which we have suffered.

"Perhaps fuel and speed restrictions imposed for reasons of economy after the World War have caused us to lose sight of the more severe demands that battle conditions would bring. Possibly our upkeep has gotten a bit perfunctory or slack. We must not let the recently established 5-5-3 armament ratio lull us into a false sense of security. The German Kaiser thought he could conquer the world. There may be others with the same idea."

Commander Donald then took over.

"As you know, gentlemen, the fuel and speed restrictions have been relaxed somewhat in this new fiscal year. This will permit us to engage in a few more high-speed exercises and more closely simulate war conditions than we could during the days of fuel shortages.

"The standard speed of twenty knots, which has been set for tomorrow's run to San Diego, will crowd our cruising turbines to the limit of their power. This order comes without warning. And it leaves no room for excuses. To be sure, you have had little time for more than day-to-day maintenance since the Fleet Exercises began and, actually, your entire plants would have to be in top-notch condition to maintain the power that is required. I know what you are thinking. But such is a sailor's lot.

"On the other hand, Squadron 11 is no worse off than Squadron 12, and if they can make a good run, as they say they will, we can make one that's better. Just remember that it is trials such as the one coming up, with no time for preparations and fine tuning, that make ships and squadrons stand out above others in maintenance and resourcefulness. It's been said before, but I'll say it again, this run will be something that, among engineers, separates the men from the boys.

"If the ships of Squadron 11 make good scores on this run and keep on pitching during the rest of the fiscal year, I expect to see the big Red E, for excellence in engineering, blossom on one of our smoke-stacks come next July. I'm fully convinced that you have the experience, the know-how, and the determination to do it."

After a few technical questions and some discussion, Captain Watson dismissed his officers with a friendly graciousness that marked him as the leader that he was. Captain Watson's quiet yet resolute manner had won him many friends throughout the Navy.

"That will be all, gentlemen. Good luck, and as our Canadian cousins would say, God bless."

ComDesRon 11 was on the threshold of 50, a true salt-horse sailor with graying hair and a close-cropped, somewhat bristly, mustache. Of stature, he was medium tall, a bit heavy-set but in fine physical trim, and he bore himself with an air of distinction, authority, and decision. As an officer, he was genial but exacting. His voice was soft. On the other hand, he was rather quick-spoken for a son of the Bluegrass State. Edward Howe Watson was not only the scion of an old Kentucky family, but a Navy Junior to boot. His father, who retired as a Rear Admiral in 1904, was a member of the Naval Academy Class of 1860 when that institution was but 20 years old.

John Crittenden Watson served under Admiral Farragut during the battles of New Orleans and Mobile Bay. Virtually, from the day he was born in 1874, Edward Watson had been destined to become a Kentucky gentleman and an officer in the United States Navy. He succeeded admirably in both respects. After being graduated from the Naval Academy in 1895, Edward Watson served in various commands. His record was so excellent that, in 1913, he was selected for the two-year course at the Naval War College.

On completing it, he served for a time as Executive Officer of the battleship Utah. In March 1918, he was given command of the battleship Alabama. This he skippered during World War I in Atlantic waters and won a citation for meritorious services. He was detached from this command in January 1919. Following extensive briefing in the Office of Naval Intelligence, Captain Watson was named U.S. Naval Attaché at the American Embassy at Tokyo, Japan.

This, it will be remembered, was a period when the international barometer was very low in the Pacific. He remained there until the spring of 1922. Then he returned to this country to take command of Squadron 11, Destroyer Squadrons, Battle Fleet, and hoisted his Commodore's flag in the Delphy. During the years of his retirement, Rear Admiral Watson had followed, with pride and interest, his son's rise in the service they both loved so well.

In 1923, what with the promotional log-jam that followed World War I, advancement to flag rank was not easy. But old-timers in the Navy took it quite for granted that, among those who had their fingers firmly on their numbers, Eddie Watson would be sure to make Admiral. He had the training, the record, and the capacity to assume responsibility. To live to see this achievement fulfilled was the long-cherished dream of Rear Admiral Watson, then more than 80 years old.

He was to die within three months of that September day with the sad knowledge that his fondest hope regarding his son would never be realized.

The four-stackers of World War I design were, in 1923, the pride of the Navy. Compared with the high-speed, heavy-hitting power and navigational aids of destroyers of today, they do not measure up impressively. But in this story we are looking back almost 40 years at the old DDs. And let us not forget that, old though they were, 50 of them through the operation of Lend-Lease, were good enough to help save England in 1940-41.

So far as equipment and performance were concerned, the long, lean combat greyhound was the Buck Rogers vessel of the War to Save the World for Democracy. Her maximum speed was phenomenal. Her armament—guns, torpedoes, anti-aircraft and anti-submarine weapons—was powerful. A veritable dreamboat of destruction in the eyes of the Cods of War.

From knife-sharp bow to elegantly rounded fantail, the typical destroyer in Squadron 11 had a length of 314 feet, a beam of 32 feet, a displacement of 1250 tons, and a draft of nearly 10 feet. She had two high-power and two low-power turbines that could deliver no less than 27,000 horsepower.

Her two triple-bladed propellers could push her through the seas at 32 knots or at a rate of approximately 36.9 land miles per hour. On her main deck—flush from stem to stern—stood the four rakishly planted smoke-stacks that served her four high-pressure boilers. Between the #2 and #3 stacks lay the galley deckhouse where the cooks fussed over their highly important pots and pans. It was topped by two rapid-firing 4-inch guns similar to one mounted on the forecastle.

Forward of the smoke-stacks was the bridge structure with its chart house, radio shack, and Captain's emergency cabin. Aft of the stacks, in the waist of the ship, were four sets of triple torpedo tubes—two to starboard, two to port—separated by a tall, slim searchlight tower.

Next came the after deckhouse with the torpedo storeroom and crew's head. On this housing was mounted the fourth 4-inch gun. On the fantail stood the 3-inch anti-aircraft gun and a rack for the release of depth charges. For her time and kind, this seagoing hornet packed considerable sting. Two masts completed the picture. Between them, they supported the wireless and radio telephone antennas. The foremast, some 40 feet higher than the stub mast, also carried a crow's nest and a signal yardarm.

Normally, a destroyer's complement would be her Captain, 7 Wardroom Officers, 10 Petty Officers, and a crew of 114. But, due to tight postwar economies, nearly all destroyer personnel ran from 20 to 30 percent short of full strength. In many cases, ensigns performed the duties of full-fledged Lieutenants. In those days, because of undermanned conditions, the heads of Commanding Officers rested as uneasily as those of kings were wont to do.

And, to be candid, the upcoming 20-knot engineering run—all the 427 miles from San Francisco to San Diego—gave all three Division Commanders in Squadron 11, as well as the Captains of the DDs that would take part in it, good reasons for feeling a bit uncertain.

They were far from sure that their vessels could make the run at such continuous high speed—about 23 land miles per hour—on the cruising turbines.

As previously noted, in September 1923, the Navy had only recently been released from the stringent fuel economy order in force since the war years. During that dreary period destroyers had been restricted to 12 to 15 knots. As a result, their engineering plants had been insufficiently tested for higher speeds. The impending run had been advocated by the Squadron Engineer Officer, Lieutenant Commander "Blinky" Donald, because many engineering troubles had been encountered during the summer.

He felt it was essential to get a good test of the squadron's engines under maximum demands. Hence the standard speed of 20 knots ordered for the run from San Francisco to San Diego was just what he desired. This was a crucial speed because the cruising turbines could just barely make it and no ship wanted to resort to the use of her main turbines because of the resulting increased use of fuel and her correspondingly lowered standing in the engineering competition.

Destroyers, who could make it on their cruising turbines, were therefore in the best engineering condition. It was a challenge, and, while all concerned were eager to meet it, they were none too sure about the ability of their individual ships to make the grade without a bit more preparation and working up.

But, as one might say, the order of the day called for a cheery "Aye, aye, sir" and a brisk "Let's go, sailors. What are we waiting for? Squeeze the last drop of horsepower out of those turbines. Let the old girls shake and shiver till they rattle your back teeth. Let's go!"

These tin-can sailors had no fear of wind. They had no dread of seas. They were used to a coastline hidden by a mantilla of fog. But they failed to reckon with the jinns of Honda. They neglected the warnings of bilge rats that scented danger, of a Chief Petty Officer driven AWOL by nameless fear, and of a sky that glowed crimson in the dawning. But let us take a glance at that sunlit early September morning in San Francisco's traffic teeming harbor.

Bridge builders had not as yet spun their steel cables across the Golden Gate or between San Francisco and Oakland. Ferry boats, flotillas of them, scurried like clumsy water-beetles from shore to shore and scooted around the great, gray battle-wagons that rode placidly at anchor in mid-channel. Abroad the destroyers forming up in the wake of the Squadron Flagship Delphy, from bridges to boiler rooms, they were ready for proceeding to sea and for whatever signal from the Commodore might speed up their forced draft blowers and send them racing to new stations for the morning's maneuvers.

Destroyer Squadron 11 did, indeed, make an inspiring picture as it moved slowly past the San Francisco waterfront. Colorful flag hoists, identifying each ship in the International Code, flew from their yardarms. Signal lamps blinked messages from ship to ship.

Leadsmen in the chains called their soundings. Captains and navigators were at the conn. The national ensign (steaming colors, to destroyer men) flew at the gaff of each cocky-looking stub mast. Many a heart that morning beat faster with pride and gratitude at the sight. Gathering speed, with the Commodore's broad command pennant flying at her fore, the Delphy, trailed by her slim, gray teammates, stood out the channel for the appointed rendezvous off San Francisco Lightship.

Grim Alcatraz Island, the beautiful Golden Gate, and the rough, shallow Potato Patch outside were quickly passed.

Gracefully, then, as though in a curtsey, she and her brood dipped their bows deep into the long ground swells of the mighty Pacific.

The rays of the sun danced blithely on the foam-flecked waters and highlighted immaculate metal and paintwork on the briskly steaming men-o'-war.

No one could know that seven of these gallant little fighting ships would never enter the Golden Gate or any other port again.

USS SAN FRANCISCO UNDER THE GOLDEN GATE BRIDGE, 1942

3 – Honda: Rugged and Untamable

On that September 8 in 1923, Honda, as always, waited with hopeful and hungry patience for offerings brought by the jinns of wind, wave, current, and fog. It stood as changeless in appearance as it did in character, desolate and dangerous. In the first quarter of the twentieth century, the progress and development that had come to fruition in other parts of California had not reached Honda, nor has it to this day.

Difficult to reach, except by sea, and yet offering no safe anchorage, Honda remained the bleak no man's land it had been before the days of Spanish occupation in upper California. In 1923, the mesa's sandy soil supported the same sparse growth of grass, cactus, and desert plants it nourished about a century earlier when, as part of a 24,992-acre land grant, it was ceded to Anastacio Carrillo by the then still young Mexican Republic.

The huge grant—greater in length than in width—ran about fifteen miles along the coast from Canon del Coja on the Santa Barbara Channel, northward to the deep ravine called Canada Honda. The entire region was known as Punta La Concepcion. In 1863, President Abraham Lincoln confirmed the title to the Carrillo family. In later years, the northern half of the grant passed into the hands of Captain Robert Sudden. In 1923, his son Robert, as at the time of this writing, lived with his family at his Rancho La Espada about halfway between Point Arguello and Point Conception. The former is a few miles south of Honda. The latter is the upper gatepost of the shipping lanes that lead into and out of the Santa Barbara Channel.

Thirty-seven years ago, only one rough and winding road—called the Sudden Grade—led to the Honda region from the outside world via the little town of Lompoc, up the torturous Miguelito Canyon road and over the rollercoaster country formed by the Lompoc Hills—high-rising, steep-sided heaps of rock and sand that stand like tumultuous breakers on a petrified sea.

To be sure, there were an almost impassable wagon road and a few narrow pack-trails. As for the latter, they demanded the agility and surefootedness possessed only by burros and mountain goats. In 1901, after two years of herculean labors, the Southern Pacific Railroad completed the work of constructing approximately 50 miles of single track along the narrow coastal plateau of Point Conception.

For a full quarter of a century Honda had played its role in holding the Southern Pacific at bay. It had, in fact, because of the engineering difficulties involved, contributed its part toward keeping the coastal railroad system divided into two halves. The northern half of rail began at San Francisco and ended at Surf. The latter is a small fishing hamlet that sends a sidetrack to Lompoc some 9 miles inland. About 50 miles to the south of Surf, at the village of Elwood on the Santa Barbara Channel, the rails resumed their reach toward Los Angeles. From Point Conception, the coast trends in a gentle northwestward curve for about 12 miles to Point Arguello. Next it runs northeastward, in another gentle curve, a good 10 miles to Point Purisima. The only things that are gentle about this stretch of ship-killing coastline are the two curves just mentioned.

The railroad builders found the terrain between Point Arguello and Point Conception especially difficult to tackle. The coast consists of a series of bold, rocky cliffs, 100 to 400 feet high. They had to lay their rails along the slopes of these obstacles or drill and dynamite their way to build tunnels.

Between Surf and Honda—a short but tough 5-mile run—the engineers were called upon to throw tall trestles across the yawning chasms of three major ravines gouged out of the mountains by the sharp chisels of flash floods.

First came the wide gap of Bear Canyon, next the narrower Spring Canyon. Last to be bridged was the broad engineering jump across Canada Honda, with its usually dry creek bed snaking through the bottom of the defile far below. There were other but minor gullies to span in this formidable stretch of coastal wilderness, but Honda, Bear, and Spring Canyons comprised what the railroaders came to call the Terrible Three.

To their left and ahead, as construction gangs worked southward from Surf and up a slowly rising grade toward Honda mesa, towered the sharply ridged ranges of the Santa Ynez Mountains. On their right, and increasingly farther below, the great swells of the Pacific rose with deceptive laziness into breakers that crashed thunderously as they assaulted the rocks and pinnacles that form an almost constant barrier between sea and shore.

On the completion of the railroad link between the two halves of the Southern Pacific coastline, human dwellers—in the guise of a 16-man railroad section crew—came to Honda mesa to remain only a few decades. To ensure the continued safety of the track, a section gang was established along the rails on the mesa, a quarter of a mile south of the trestle that crosses Canada Honda and a good quarter of a mile up the slope from the cliffs, rocks, and reefs variously known to men of the sea as the Devil's Jaw, Point Pedernales, Graveyard of Ships, or just plain Honda.

Because of the prominent roles they were to play in the impending tragedy at Honda, the stretch of rails between Surf and Honda mesa, as well as the quarters of the section gang, must be brought into full focus. This applies also to Section Foreman John Giorvas, a square-set man of Greek ancestry, in early middle age, with a pugnacious jaw and quick brown eyes. He lived in a story-and-a-half frame house that contained a living room, a kitchen, and a storeroom on the ground floor and two small bedrooms upstairs.

Not large quarters, but they suited John Giorvas well enough. The laborers in his section gang, all Mexicans, lived in two long, low bunk-houses that were divided into single rooms, each with its own entrance. Between the bunkhouses was a barn that sheltered the section's hand-driven track car. Like all Southern Pacific structures of that era, the buildings were painted a dark yellow with brown trimmings.

A double row of trees, planted along the line of buildings in 1901, had by 1923 grown tall and stately. Drinking water, an extremely scarce commodity in this vicinity, came from a subterranean reservoir by means of a solitary pump. The men, the buildings, and the pump have gone but the trees still bend stubbornly before summer winds and winter gales.

Up the track toward Surf, near the northern end of the Canada Honda trestle and half a mile from the small group of section gang buildings, stood a small frame building hardly larger than a toolshed. This was Honda Station. But it rated only as a seldom-used flag stop.

There was neither station master, ticket seller, freight handler, nor telegrapher. In fact, Honda's only means of swift communication with the outside world was a telephone line that ran from an instrument in John Giorvas' living room to the

Southern Pacific dispatcher's office in Surf, exactly 5.8 miles away.

To be sure, John had an old-fashioned wall phone that connected him with a party line of the Santa Barbara Telephone Company. But this outlet was usually clogged by the long-winded conversations of ranch wives and/or their teenage offspring. Giorvas felt that he should be paid time-and-a-half whenever he tried to put a call through on that phone between dawn and bedtime. Giorvas was a somewhat junior foreman on the Southern Pacific's roster when he first came to Honda in the late 1910's.

It was a far from desirable assignment along the railroad's coastline system, but there was something in the nature of Honda and in the make-up of Giorvas that complemented each other. The years went by. Section hands came and went in a fairly steady procession, but John Giorvas stayed on.

As time passed, he learned to understand the eternal melancholy of the mesa with its backdrop of changeless mountains. He also came to know the unpredictable moods of the dictatorial sea. To his lively imagination, the sheer black cliffs and wave-honed rocks of La Guijada del Diablo were sources of constant fascination.

On Sundays, or after work on weekdays, Giorvas would often pick his way down the sloping mesa to the edge of the bluff. That is, except when thick curtains of fog blotted out all vision over land and sea. And that was rather frequently. In fact, over long periods of time, the keepers of nearby Point Arguello Light annually entered more than 2000 hours of fog, year following year, in their watch logs.

The steep wall of volcanic rock that runs along this section of the California coast bulges seaward in a fairly well-rounded curve west and slightly north of the section gang's buildings. At the same time it rises sharply from a level of some 30 feet to one of about 70.

This elevation is further heightened by a sharply rising upslope near the northern end of the bluff that ends in a small, almost circular, hilltop some 300 feet above sea level.

In 1923, this point of elevation was one of a brace of twin peaks connected by a sway-backed strip of land which gave the peaks the name of Saddle Rock. The horn of the saddle was the peak just back of the bulge of volcanic rock. Officially, on charts and maps, the site was known as Point Pedernales. The name, which means "flints," derives from the large number of arrowheads, fashioned from flint by Indians, and found on the site by rancheros in the days of Carrillo ownership.

To the left of the bluff, and only about 30 feet from shore, is a smallish island with precipitous sides that rise about 25 feet above sea level. The island—it has no name, but let us call it Bridge Rock—is about 100 feet wide and twice as long. Its surface, cut into a crazy pattern of keenly edged crevices, offers at best poor footing. Jagged rocks rise here and there. Many of them are needle-pointed and sharp as Mafia stilettos.

A peculiar feature of this nameless hunk of lava is that, despite appearances, it is not an island at all but is connected with the shore by a narrow, rocky bridge. Built by nature, this link to the mainland is barely 3 feet wide and more than 20 feet above the seas that surge impatiently beneath it. This dangerous avenue of escape from the wave-battered rock was to serve as a major prop on the stage of disaster during the action of the Tragedy at Honda.

Surrounding this island-like projection, but most of them submerged even at low tide, are numerous rocks and ledges. In a broken chain they run seaward for a distance of 500 yards, where they surface in the form of two large masses of fang-like pinnacle rocks surrounded by smaller but equally formidable satellites. So much for the left side of the Devil's Jaw.

The right extension pushes seaward from close inshore, off a sandy beach formed by Honda Creek, in the form of a series of visible or submerged pinnacle rocks and ledges. It was here the liner Santa Rosa struck in 1911 with great loss of life and complete loss of the ship.

Twelve years later, the rusted engines and broken boilers of the luxury steamer still stood as somber tomb markers just above the narrow sandy beach.

Even when the sea is smooth, lively surgings of surf and swells sweep continually over the exposed pinnacle rocks as if to keep them well whetted against the time when they will be called upon to rip through the bottoms or gash the sides of luckless vessels.

In the entire reach along the New World's western shores from Tierra del Fuego to Alaska's Point Barrow there is hardly a coastline that offers a greater concentration of maritime dangers than the strip which runs between Point Purisima and Point Conception.

The latter, even today, is labeled in the United States Coast Pilot as the Cape Horn of the Pacific. This because of the heavy northwesterly gales encountered off it at the turn into Santa Barbara Channel. As an active menace to this vital gateway lies San Miguel Island, 23 miles due south of Point Conception. This desolate pile of lava towers, in places, to more than 800 feet.

Its bold, broken, and rocky shores are guarded by dangerous reefs, sunken rocks, and great boulders that stand awash even at high tide. San Miguel is a place mariners avoid as if this dark and deadly island were the Black Plague itself. It was that way in the olden days. It is that way today. From Point Conception north to Point Purisima, the rugged shore lies behind a virtually continuous barricade of rocks, reefs, and shoals that has earned this line of coast the grim title of Graveyard of Ships. And, virtually in the very center of this cemetery of the sea, like a black widow spider waiting patiently for victims in the middle of her web, stands Honda topped by Saddle Rock, which John Giorvas used as his place of observation and contemplation.

Here he would sit, hours on end, listening to the orchestra of the ocean: the reeds, horns, and strings of the winds; the measured trap-drums of the surf; the kettledrums of the breakers. From this point of vantage, he had not only a gallery seat that gave him an excellent view of the oceanic stage before him but also a full sweep of the shorelines that run into the distance on either hand. Maritime traffic was heavy in those days. Highway engineering and automobile design had not yet combined to create the Motor Age which moved transportation in Pacific coast states from keels to wheels, from seaways to highways.

Now and then, sailing vessels—even an occasional square-rigger—would glide down along the coast or beat up against the wind. From his lookout, Giorvas would watch wooden steam-schooners, lumbering freighters, graceful liners, or gray-coated men-o'-war stand down the coast for Santa Barbara Channel or come into view from the south after rounding Point Arguello.

To be sure, nearly all of them—tramp or trawler, lumberman or liner, or man-o'-war—gave Giorvas' vantage point a wide berth. Honda is not the sort of place for which even long-standing familiarity breeds the slightest contempt. The pathetic remnants of many wrecks give ample proof that the Devil's Jaw is indeed a graveyard where many small and large ships have come to their final port o' call. The list of vessels that came to grief in this region is too long and it covers too great a span of time to warrant presentation here.

Still, it is interesting that one of the first steamers to round the Horn under the Stars and Stripes, after the start of California's Gold Rush in 1848, was the Edith. She ran aground off Honda on a clear night. Some put the blame on a treacherous current with a northeast or landward set. Others claim that the gold-crazed crewmen deliberately beached the steamer and piled ashore without further thought for the welfare or the safety of their ship, in order to make their way to the gold country. If this latter surmise is correct, chances are that none of them realized the long, hard journey they had ahead of them.

To gold-seekers, California meant but one thing—fabulous riches. Six years later, on the afternoon of September 30, 1854, Captain Henry Randall, in the pilothouse of the luxury steamer Yankee Blade, gave the departure signal. With hoarse blasts bellowing from her whistle and the ship's huge paddle wheels churning slowly astern, she glided smoothly away from her San Francisco pier. Crowds ashore and passengers aboard cheered, shouted, and waved enthusiastic farewells.

In her elegant First Class, modest Second Class, and miserable Steerage, the Yankee Blade carried a record load of passengers. Many of these were miners returning to their eastern homes with heavy money belts; but even greater numbers were made up by a miscellaneous lot of disappointed gold-rushers who somehow or other had begged, borrowed, or stolen enough money to pay their fares back to the Atlantic States by way of Panama. Amazingly enough, there were quite a few women aboard and some of them had children. Amidships, in the vessel's gold room, which was below the passenger decks, was stored $153,000 worth of the yellow stuff. It was being shipped by Page, Bacon & Co., San Francisco bankers, to one of the great banking houses in Philadelphia.

The vessel's complement consisted of Captain Randall, his Deck and Engineering staffs, and a crew of 122 men. Twenty-four hours after her departure from San Francisco, as she was standing southward along the California coast, the Yankee Blade ran into a dense fog. Captain Randall could have met this situation by reducing his speed and heaving his lead. But, believing himself to be safely at sea, he let his paddle wheels keep churning and sounded intermittent fog warnings with his whistle. What Captain Randall did not know was that, somewhere along the route, he had met up with a wayward, contrary current with a strong inshore set.

Instead of being to the west of San Miguel and heading due south on his course to Panama, he was actually off Honda and being set to the eastward.

Just as the Quartermaster of the Watch struck 8 bells at the beginning of the First Dog Watch—4 P.M. to a landsman—there was a terrifying grinding, tearing sound. Personnel and passengers were thrown off their feet, while loose gear, baggage, and galley equipment crashed to the deck as the once-proud vessel came to an agonized and humiliating stop. Her decks slanted sharply upward and her bow stood high and dry. She had gone the way of other ships that had run afoul of Honda and its rocks. Because of her speed, the vessel had run more than 50 feet onto a rocky reef on the southward or Point Arguello side of Honda's Bridge Rock.

There she hung, with her stern suspended in 9 fathoms of water. Her bottom had been badly ripped and her sides heavily slashed by the sharp fangs of the Devil's Jaw. Thanks to the coolness of officers and deck-force, the pandemonium that threatened did not break out. Women and children were directed to assemble at the boat davits. The men of the fireroom watch lifted safety valves to prevent boiler explosions. Engine rooms and boiler rooms were vacated as the ship settled steadily and rapidly at the stern.

The First Officer was in charge of the lifeboat on the starboard quarter. This craft, while still suspended to the davits and flush with the deck, was filled with women and a sprinkling of men.

It was being lowered to the sea when the after-tackle was let go. The boat swung by the bow as its human cargo was dumped into the sea. Several women were drowned, others were saved by lines thrown to them from the deck. The men—mostly members of the fireroom's black gang—grabbed the gunwales of the lifeboat, whose forward falls had now been released, as it floated, awash, on the heaving swells.

Meanwhile, as if dissolved by the jinns now that it had served its purpose, the fog lifted abruptly. A few hundred yards away and sharply outlined were the ugly volcanic cliffs of Honda and its slowly rising mesa. The First Officer's boat eventually headed shoreward, only to be swamped by breakers. A few minutes later another lifeboat, with about 30 souls aboard, suffered a similar fate. Many of the men in this boat were dragged down to their death by the weight of their gold-belts. Sixteen men and women were drowned. Watching these disastrous landings from the deck, Captain Randall decided that it was more important for him to get passengers ashore alive than for the Captain, in accordance with the tradition of the sea, to be the last man off the ship.

Another boat was launched, with Randall at the tiller.

By dint of superb seamanship and the grace of God, he brought his boat through the breakers and reached safety on the small sandy beach that lay in the lee of Bridge Rock. Soon another boat, handled by the Third Officer, landed her passengers in the cove.

For the next half-dozen hours, until returning fog and the mutinous attitude of the exhausted men at the oars brought a stop to the rescue operations, the two boats brought about 150 passengers to the dubious security of Honda mesa. Meanwhile all hell had broken out aboard the Yankee Blade. With the officers gone, a gang of plug-uglies emerged from the Steerage with the boldness of rats from the bilge. Pointing six-shooters and brandishing knives, they spread terror among the passengers and cowed the crew.

Despite the fact that the ship was settling at the stern, and could break in two at any minute, they swept through the First Class after-cabins, where they ransacked hand baggage and trunks. They had to work fast.

And they did. Within 30 minutes after the Yankee Blade struck, the stern had sunk to the level of the promenade deck. To their great disappointment, the hooligans could not get at the gold shipment or the valuables deposited by passengers in the Chief Purser's vault. Both were under water and beyond reach.

The situation steadied somewhat when Captain Randall and his Third Officer—the First and Second Officers had vanished without trace—returned to the ship about 10 P.M. Most of the passengers spent a relatively peaceful night crowded like sardines into one of the forward lounges.

Some sang hymns; others offered softly spoken or loudly shouted prayers; still others sweated out the suspense of the slowly dragging hours in mute silence. From below rose the ribald yells, lusty swearing, and coarse laughter of the ruffians.

They had broken into the ship's liquor stores and were having a high time drinking whatever was at hand. And now and then, when spells of unaccountable silence swept over the ship, one could hear the doomsday crash of nearby breakers as well as the mournful tolling of the Yankee Blade's bell as the clapper swung back and forth in keeping with the vessel's wave-rocked motion.

Ashore, conditions were almost equally intolerable. While their actual physical danger there did not compare with the potential dangers that threatened those aboard the stricken ship, the discomforts and threats of violence were considerable. Most of the 150 passengers who sought shelter from the cold night wind on the inhospitable mesa were women and children. But among the men who had come ashore were quite a few of the looters. Being armed, the "hellhounds," as they were called, soon had the situation well in hand. They boldly helped themselves to coats, blankets, canvas, and other coverings badly needed by the water-soaked, shivering women who vainly sought to start and maintain fires with the stunted brush that grew so sparsely upon the desert slope.

The only way these unfortunates could recover their belongings was to pay for them with whatever money or valuables they had on their persons. There was neither food nor water nor medical relief of any sort for the many who had suffered cuts and bruises in climbing the steep cliffs from the sea to the mesa.

With the coming of dawn on October 2, rescue operations were resumed by Captain Randall and his Third Officer. At the same time, stores of food and other necessities were loaded into the boats. Casks of water, light enough to float, were heaved over the side. Many of these were splintered on the rocks, but quite a few made the shore and were recovered.

By noon, there were signs that the Yankee Blade might break up at any moment. At that time, less than half of the hundreds of passengers aboard had been ferried to terra firma. To Randall, it seemed an impossible task to take the remainder ashore.

Just as the hopelessness of the situation began to wear him down, relief came in the appearance of the S. S. Goliath, commanded by Captain Thomas Haley, a coastwise packet that ran between San Francisco and San Diego. The southbound Goliath picked up those who were still aboard the Yankee Blade and continued her journey. Before she left, Captain Haley lent Captain Randall a dozen or so six-shooters and an ample supply of ammunition.

Ashore, the guns were distributed among responsible male members of the party and order was restored, at least on the surface. Records, however, reveal that during the first two days there was a constant black-market trade in goods taken from the wreck or in the sale of clothing and other belongings the bully-boys had taken, by force or otherwise, from their owners.

Meanwhile, Captain Randall directed his Third Officer to head a search party for assistance. After stumbling many weary miles up and down the coast, as well as inland toward Mount Tranquillon, the searchers returned without having found any trace of human habitation. Although it was late in the season and almost time for the southeasterly gales that howl over this region from fall to spring, the weather held.

The days dragged by. Not a ship was sighted close enough to be hailed. Not a sign of rescue at hand. The Yankee Blade, with the pounding of the surf, had broken in half. Nothing remained of her stern but a shell that had turned bottom up. Her forward portion, showing her two tall smoke-stacks, had slid down the reef but was still kept upright in the cradle of rocks that had held her fast.

Rescue came on October 8 when the Goliath returned. Captain Haley took a calculated risk and brought his vessel as near the dangerous shore as he dared. Through smart handling, lifeboats soon had every one off Honda's inhospitable shelter and aboard the rescue vessel. Thus ended the saga of the sinking of the Yankee Blade, except for one small postscript. In 1856, the Santa Barbara Gazette reported that a Captain Randall, skipper of the schooner Ada, had succeeded—where many others had failed—in recovering all the gold from the wreck of the Yankee Blade. His share of the salvage, so said the newspaper, would be "upwards of $80,000." Could that have been our doughty Captain Harry?

In the course of the next 50 years, there were many tragic but largely undramatic and unimportant—if any disaster at sea can be called unimportant—strandings of smallish vessels that left their hulls to rot or rust in the Graveyard of Ships.

The jinns of Honda did not capture a really outstanding victim until July 7, 1911, when the passenger steamer Santa Rosa ran aground just off the slowly shelving half-moon beach formed by the creek waters that occasionally run from the mountains to the sea via Canada Honda. This gully lies a quarter of a mile north of Point Pedernales.

The stranding happened during the small hours of the night. There had been fog, but by the time the hapless vessel was sighted by Keeper W. A. Henderson of Point Arguello Light, a light breeze blew from the northwest. This benevolent wind brushed the fog away without disturbing the relatively smooth surface of the sea beyond the lines where ocean swells change into the breakers that form the eternally pounding surf.

When Henderson and two assistants arrived on the scene about 5:30 A.M., the Santa Rosa was lying broadside to the beach and resting easy. If help came before the tide got too low, it should not be too difficult to get her off. Through some major miracle, the Santa Rosa had not been ripped by a single fang in the Devil's Jaw. Soon after Henderson arrived, the steam schooner Helen Drew hove into sight. Almost simultaneously the mail steamer Argyll appeared on the scene. Both vessels passed lines to the stranded steamer. The lines snapped like thread. Next the Argyll made fast a wire cable. However, by that time the tide was ebbing and nothing could be done in the way of towing her off the beach until the next tide.

If wind, wave, and weather held until the afternoon high tide, there was still hope for the Santa Rosa. But these hopes began to fade about noon when Honda slowly but effectively set about to live up to its evil reputation. By then the wind had gained considerable strength. The surf was rolling high and breaking hard. At 4 P.M., the Argyll again put strain on her cable.

The Santa Rosa's hull lay north and south. The Argyll's pull was westward into deeper water. Slowly the Santa Rosa swung around. She had almost made it when the Argyll's cable broke under the strain. In moments, the great swells had seized and broached the Santa Rosa. In this most critical instant, when everything hung in the balance, the jinns of Honda played one of their trump cards. A tremendous sea, a veritable wall of white-capped green water, thundered shoreward. It listed the helpless Santa Rosa as if she had been a child's toy, flung her against submerged rocks, and broke the vessel's back amidships.

For reasons that have never been recorded, the Argyll did not stand by. At any rate, the Santa Rosa was left to her own devices and those aboard her had to save their lives as best they could. All that Henderson and his assistants could do as, fascinated with terror, they stood upon the beach, was to watch the stark drama that now was enacted. First, a lifeboat came around the bow of the steamer, evidently for the purpose of passing a line ashore. But a huge comber capsized it and left its crew struggling helplessly in the surging seas.

Their drowning cries were heart-rending. One man was seen to be flung toward the beach, where he was alternately hurled shoreward by breakers and dragged to seaward by the undertow. Henderson and his men hurriedly grasped hands to form a human chain, and dashed into the breakers. Twice the chain was torn apart, but, on a third try, the man, more dead than alive, was snatched from the jinns of Honda.

Meanwhile, the cries of his unfortunate fellow crewmen had been silenced. They were dead. Following this disastrous start, a raft and a lifeboat were lowered over the ship's side. They were filled with passengers and cut adrift. An extra-large comber rolled in. As the raft began to ride it, the wall of water broke, tilted the raft as though it were a chip, and dumped its doomed occupants into the crashing surf. The lifeboat had better luck.

A heavy breaker seized it much the same as a football player grabs a ball, makes a run, and scores a goal. This one streaked toward shore, lifted the lifeboat and, with mighty force, threw it high upon the beach.

Soon after that, a line was shot toward shore. But something went wrong; the line parted. Next, another boat was lowered. It edged inshore to a point where a man in the bow could throw

Henderson a heaving-line. The boat returned to the ship, made fast a hawser and a lighter-line which were hauled ashore. But where to fasten the hawser which evidently was intended to support a breeches buoy? By now, a few men of the Southern Pacific's section gang at Honda had arrived on the scene from a job up the track.

Their foreman had brought along stout cable and tackle. He saw that the only way to rig a trolley for the Santa Rosa's breeches buoy would be to secure its shoreward end to the girders of the railroad trestle where it spanned Honda Canyon.

True, the trestle was some distance from the shore, but, if the Southern Pacific ropes were bent on to hawsers from the stranded ship, their combined lengths would be sufficient. This was done, and, with block and tackle, eager hands hauled the overhead line taut and secured it firmly.

Having no regular breeches buoy, and deeming a boson's chair too light for the rugged journey, officers of the steamer first used a large, heavy oil drum as a carrier. But this was so heavy that, when it came halfway in, the hawser sagged and the drum dipped into the surf. There it was simultaneously kicked about by the fierce wave action and filled with water. Although all hands on the beach, women as well as men, hauled with all their might, it looked as if the passengers in the drum would either be spilled into the surf or drowned in their container before they could be got ashore. Eventually, the rescuers won.

The next time, a cargo net was made to serve as a carrier. Rough though the treatment was for its passengers, the net saved numerous women and children—even some tiny babies. From then on, the cargo net was shuttled with success, back and forth between the ship and the shore.

For those aboard the Santa Rosa, the wreck and rescue was a near brush with death. For those on shore, it was back-breaking and exhausting work. Henderson and his few able-bodied helpers were near complete collapse, but they hung doggedly on. Things ran a little more smoothly when crewmen began to arrive in the cargo net and the rest of the hands in the section gang showed up. By sunset a raft was put into service to serve as a rope ferry.

At 10:30 the last passenger had been taken off the ship. But the work was not over. Until long past midnight, men strained at the task of bringing food, drink, shelter, and other necessities for the weak and weary unfortunates who were crowded upon the beach. Soon after daylight, a special Southern Pacific train arrived at Honda Station to pick up the survivors.

As for the Santo Rosa, within a few days the violence of the surf had widened the crack she had received amidships to a point where she broke into two. There was nothing left to salvage.

In the fall of 1923, a full 12 years after the Santa Rosa stranding, the Santa Rosa's broken engines and rusted boilers still showed above the seething water and lay, half buried in the sand, just north of the little cove into which Honda Creek empties. From his favorite seat atop Saddle Rock, John Giorvas, now foreman of the Honda section gang, could see these pathetic relics of a once mighty ship. The sight of them always turned his thoughts to the tremendous power of Mother Nature's elements and the frailty of Man's greatness.

Honda!

There had been no sinkings in recent years. What with the railroad track stretching its gleaming threads along the mesa— the long-drawn hoot of locomotive whistles and the rhythmic roar of wheels rolling over the rails—it seemed as if the jinns of Honda had lost their power to perform black magic with wind, wave, fog, and current as was their wont during Honda's era of absolute desolation before the turn of the twentieth century.

At any rate, in 1923, a full dozen years had floated down the currents of time since the jinns had executed their last major coup and claimed, for the diabolical sea, its last major victim.

FLEET WEEK, SAN FRANCISCO, 1908

4 – PRELUDE TO DISASTER

Cavalry of the Sea

"EXECUTE COURSE PENNANT SIGNAL," Commodore Watson, standing in the wing of the navigating bridge, called quietly, as the Delphy, with a high-curling bone in her teeth, raced past the San Francisco Lightship at 20 knots. The Quartermaster of the Watch had just struck one bell—0830—of that fateful Saturday morning of September 8. At his order, repeated to the signalmen by the ODD, the Delphy's flag hoist, "160 Corpen," came down on the run, as did the answering hoists flying at the yardarms of all the 14 destroyers speeding along in her wake out from the Golden Gate toward the blue-gray Farallones.

Alert signalmen shouted, "Execute Corpen, sir," to their captains, helms were put over and decks slanted sharply to port. DesRon 11 swung gracefully to the new southerly course of 160 degrees, true, toward that distant San Diego, whose narrow entrance, dominated by the bold headland of Point Loma, they expected to reach at the crack of dawn the following morning.

The Commodore braced himself against the slope of the deck and, from his vantage point, watched with pride the accuracy of their turns and the smartness of their appearance. In spite of a tough summer in a climate where weather suitable for painting was hard to come by, his sleek chargers had managed to keep themselves well groomed.

Occasional early-rising salmon fishermen raised a hand in friendly and envious salutes as their slow, trolling craft wallowed in the wash of the passing four-pipers. Even the station-bound Lightship seemed to rear and strain at her anchor chain as though eager to snap her bonds and join the "Cavalry of the Sea" as they charged by, homeward bound.

"Captain Hunter."

"Yes, sir."

"Make signal to form line of divisions, 1600 yards interval. Put Roper's Division in the center to start with and alert all ships to stand by for Short Range Battle Practice rehearsals. Standard speed twenty knots."

"Aye, aye, sir."

A few quick orders to the signalmen and strings of gayly colored bunting raced up to the Delphy's yardarms swaying against the sky 60 feet above her bridge. Answering hoists appeared almost instantly on the other destroyers. Inter-ship competition in signaling is always keen and the action on the bridge when flag signals are being repeated, is most interesting to watch.

The keyed-up signalmen, faces alive with interest, are intent on getting their hoist all the way up to the yardarm, "two blocked" in Navy parlance, as soon as or even before the Flagship. The Chief Signalman, crouched, with his long-glass steadied across the rail, spots the leader's flags as they are being hooked onto the halyards. If the leader does not have a snappy signal watch, sometimes, to the delight of all hands, she is beaten to the yardarm.

Among the flag signals required to carry out the Commodore's order to form what was called "Squadron Cruising Formation #5," was one of particular interest to us—and of great utility in destroyer tactics. Cruising formation #5 is one in which the individual divisions are formed in column with their flagboats in line. The Squadron Flagship is one space ahead of the middle column and, of course, is the guide for the entire formation.

Thus, when Captain Watson's order had been carried out, the disposition would look like this:

F (Squad. 11)
F (Div. 33)
F (Div. 32)
F (Div. 31)
1!!

Normally, there were six DD's in each division but, as we know, on this occasion there were some absentees. When rehearsing for SRBP, the interval between columns is 1600 yards and the distance between ships, from foremast to foremast, is 250 yards. To reach the stations assigned to their respective divisions in the shortest possible time, the commanders of the wing divisions hoisted the old familiar "Sail Baker" signal and rang up full speed. To destroyers, in peace as in war, there is no time for fancy maneuvers and simultaneous ship movements. Speeds are high and time is of the essence.

Therefore the simplest thing to do was to "two block" the well-known "Follow movements of this vessel" signal, "Sail Baker," and "pour on the coal." So often was this technique used that it had come to be the veritable slogan of the "follow the leader" school of destroyer tactics.

This doctrine, if doctrine it could be called, had often been criticized, and, just a few days later, would be excoriated by the press, viewed with distrust by the general public, and severely censured by high naval authorities. True: this doctrine did not, in the piping times of peace, always fulfill the requirements of the Navy Regulations that every Commanding Officer was responsible for the navigation of his own ship and that it was his duty to warn the OTC (Officer in Tactical Command) when, in his opinion, the course set by that officer was "leading his ship into danger."

But danger, it must be remembered, is, to destroyer men—the Cavalry of the Sea—purely a relative term. Danger is their business. Keen judgment and daring are routine components of their daily life. Faith in their leader, born of his known ability, is a prime prerequisite to swift battle maneuvers. Topping off all these requirements, the destroyer captain must have initiative, determination, and confidence in himself and in his ship. Truly, such men must be and are of a breed apart, for upon the possession and proper application of sterling fighting qualities can depend the lives of the men and ships entrusted to them—and the turn of battles. The normal employment of destroyers in war, such as anti-submarine work, escorting convoys, screening the battle line and delivering torpedo attacks—usually under cover of darkness—is dangerous enough. But to their lot also fall hopeless missions such as the one they so gallantly executed in the battle off Samar in October 1944.

There, three DDs and four DEs launched a suicidal daylight torpedo attack upon the leaders of Admiral Kurita's battleships and cruisers in a desperate attempt to save six CVEs ("jeep" carriers) from destruction. Miraculously, one DD and three DEs came back and, by their action with guns, torpedoes and smokescreens, four carriers were saved. Then there was the forlorn-hope daylight attack of five British destroyers of about the same vintage as those of DesRon 11, sent out in February 1942, to head off the German warships Scharnhorst, Gneisenau, and

Prinz Eugen, which were escaping up the English Channel from their bomb-blasted refuge in Brest. Five over-age World War I destroyers against two battle cruisers and a light cruiser is hardly an odds-on fight in broad daylight.

But, with true British grit and "never say die" spirit, they dashed in through a rain of enemy shells, dodging salvos like jacksnipes, and launched their torpedoes at ranges varying from 3500 to 2500 yards. As for results, the English claimed hits. However, the Germans said that the only damage was caused by a mine hit by the Scharnhorst. By the Grace of God, only one destroyer was badly damaged, the Worcester, but even she made it back to Harwich under her own power with 26 hits in her hull. Number one fireroom was flooded and the water was lapping at the fires in the other boiler room when she made the dock. Her Captain later described her condition as being so bad that "the dockyard decided the only bloody thing to do was to jack up the whistle and build another destroyer under it."

Such are the men who go down to the sea in "tin-cans." Such were the men who, confident in the ability and judgment of their leaders, followed their division flagboats at full speed to their stations for Short Range Battle Practice rehearsals on that sunny Saturday morning. Promptly at 0900, the shrill whistle of bos'n's pipes was heard from ship to ship above the rush of the seas and the roar of the forced draft blowers. Following the pipe came the hoarse calls of boatswains' mates above and below decks, "General quarters! All hands man your battle stations! General quarters!"

In a matter of seconds, after a wild scramble of organized disorder, guns, torpedo tubes, and fire-control stations were manned. As per doctrine, all guns and torpedo tubes were trained sharp on the bows to meet the oncoming enemy, ammunition hoists rumbled, and Executive Officers were reporting to their Captains: "Battle stations manned and ready, sir!"

Each destroyer had rigged miniature targets between the #4 stack and the searchlight platform, so as to give gun pointers and trainers on their opposite numbers, opportunity to sharpen their eyesight and accustom themselves to the unpredictable plunges of their spirited chargers.

With the gunnery season just beginning, the thoughts of every "Guns" or "Gun Boss," as the Gunnery Officer was known, were centered on the winning of the "Meat Ball" (Battle Efficiency Pennant) which he fervently hoped to see snapping in the breeze at his ship's foretruck. Visions of prize money and new rating badges for high-scoring guns and torpedo tubes were not-inconsiderable incentives to the keen-eyed men who manned those weapons.

Eventually, battle stations were secured and the Squadron settled down for a morning-long SRBP rehearsal. The target division slowed to 10 knots while the wing divisions alternately dropped back and then charged ahead to their targets, exercising pointer groups at the guns while the loading crews thundered at the loading machines: "Load"—"Bore clear"—"Ready One"—"Commence firing."

Muscles toughened and techniques speeded up as the 75-pound drill shells were slammed into the breeches of the loading machines. On the bridge of the Fuller, tall, blond, Viking-like Captain Dudley Seed was having a talk with his Gun Boss. "Yes, Captain," Ensign Riley Jackson was saying, "we are short-handed, and by the rules no man is permitted to serve at two guns. But I figure we can draw on the damage control parties and the black gang."

"Surely," replied the CO in his deliberate Alabama drawl.

"We'd have to do it in a scrap—or shift gun crews from side to side as they did in the days of sail."

"Too bad we lost Ensign Howard, sir," said Jackson. "He's a live wire and a keen coach for guns crews."

Seed nodded.

"You know, Captain," Riley went on, "Howard was a victim of the old Fuller tradition about painting one's stateroom."

"Oh," queried Seed, "and what is that?"

"Well, sir," said Jackson, "in the last two or three years, every time an officer has painted his stateroom, he has immediately been detached. Mr. Howard painted his cabin this week!"

"Oh, yes?" said the Exec, as he strolled up and joined them in the laugh. "And now what happens to the rest of us? We also painted the wardroom this week. Do we all get detached?"

"You youngsters sound as superstitious as my old colored mammy back in Alabama," drawled Captain Seed, with a dry grin, as he moved off to watch the next approach. Then he added, chuckling: "Maybe we'll have to abandon ship?"

Perhaps his old mammy should have taught him to keep his fingers crossed when he made such remarks. Meanwhile the coast and landmarks of northern California were fading into the haze on the port quarter. At 0958 Point Montara Light was passed abeam, then Sail Rock was checked off and disappeared into the lessening visibility. Finally, at 1130 Pigeon Point was passed only 1 mile to port, so that every navigator had a fine chance to fix his position. At 1145 Ano Nuevo Point was passed. Due to the deteriorating visibility and the falling away of the California coast to the eastward, Ano Nuevo, most navigators felt, would be the last land they would sight until Point Arguello Light, which was then some 175 miles ahead, was picked up.

And how wrong they were to be.

The wind during the day had freshened, as it usually does along the coast after about 1000 hours, blowing from the northwest quadrant with a speed of about 18 knots. By afternoon the sea had roughened, adding a moderate sea with whitecaps to the long Pacific swells coming in from the west. These conditions, which placed the wind and sea almost dead astern of the destroyers, naturally added to the steering difficulties of the ships. It followed, also quite naturally, that since the steersmen had to use more rudder to control their vessels, the speed through the water was somewhat slowed down. This fact, of course, would later be determined by the navigators by means of successive fixes of their geographical positions. It had not been apparent in the forenoon run because of the many changes of speed used during rehearsals.

The fact that the incidence of fog might prevent getting fixes by star sights or by the use of landmarks ashore did not concern anyone very much because of the radio direction finder stations which recently had been established at the Farallones, at Point Arguello, and at Imperial Beach—just below San Diego.

These new aids to navigation had been established in 1921 by the Navy Department and were manned by naval personnel. Like all innovations, they came in for their share of criticism from the older generation of seamen, and there is no denying that there were "bugs" in the operation which had to be eliminated. Situated as they were, it was not possible for a ship to get a bearing from two stations at once, thus giving the navigator a definite fix where the two lines crossed. Instead, it was necessary to get a bearing, say at 1200, then run until, say, 1300 and get another bearing. The distance run between these two bearings was then applied by the navigator, using his parallel rulers, along a line parallel to the course he was making good. Where this distance line fitted exactly, the distance between these bearing lines gave the position of the ship at the time of the second bearing.

Navigators had been using this method for centuries to determine the distance at which they would pass—or were passing—some known landmark.

But the introduction of radio made the system also available night or day, in rain, fog or any sort of reduced visibility. Admittedly, it held great promise but it was subject to errors, human and otherwise. When a ship sent a series of long dashes to an RDF (radio direction finder) ashore, the operator there centered his receiving apparatus on the sound, read off the bearing, and reported it to the ship. Naturally, the expertness of the operator entered into this result. Also the receiving apparatus had to be carefully calibrated to allow for deviation which might be occasioned by nearby electrical circuits, buildings, or other objects.

Finally, there was no way at that time of guarding against the sending out of a reciprocal bearing—that is, one just the opposite of the true bearing.

Thus, a station on a point extending well out to sea might conceivably report a northerly bearing, for instance, to a ship which was actually south of the RDF. The situation today is very different. As the science of electronics has advanced—and under the able administration of the U.S. Coast Guard—the entire West Coast, for example, is covered by powerful Loran stations which, operating in groups as "master" and "slave" stations, can cut in a ship's position hundreds of miles out in the Pacific.

For position finding closer inshore, say 150 miles, there are radio beacons for use by ships with RDF, and even distance-finding stations, which make clever use of the difference in speed between radio, sound, and sonar waves to obtain their results. Eventually, the picture, once so familiar to seamen, of a navigator and his quartermaster "shooting the sun" to obtain the ship's noon position, will be a veritable museum piece. Shortly after noon another engineering casualty occurred when the Thompson hoisted the breakdown flag and dropped out of column. Already the Percival had suffered two minor casualties not related to the question of speed, during the forenoon, but regained her station astern of the Fuller before 1000.

After rehearsal runs for SRBP were completed about 1100, the Squadron had settled down to a steady speed of 20 knots, and, although the black gangs were having their problems in keeping up the pace, all had appeared to be going well. The Thompson, however, was able to correct her difficulties and, by working up to full speed, caught up again and, at about 1730, fell in at the rear of her division.

Meanwhile the J. F. Burnes, Lieutenant Commander M. J. Foster commanding, had, at 1453, split a tube in #2 boiler and also was forced to drop out of formation. Since this was more serious and meant lighting off boilers #3 and #4 and then securing #1 fireroom, she fell hopelessly behind and never did regain station—the luckiest thing in the world for her.

Finally, at 1627 the signal "Form 18" fluttered from the Delphy's yardarm. This called for the Squadron to form column on her. Further signals specified that DesDiv 33 take station immediately astern of the Flagship with DesDiv 31 and DesDiv 32 following in that order.

The cruising order for the coming night run past Point Arguello, through Santa Barbara Channel, and on down to San Diego was to be as follows:

DesRon 11—ComDesRon, Captain Edward Howe Watson
USS Delphy—#261—DesRon 11 Flagship—Lt. Commander Donald T. Hunter

DesDiv 33—ComDesDiv, Captain Robert Morris

USS S. P. Lee—#310—DesDiv 33 Flagship—Commander William H. Toaz
USS Young—#312—Commander William L. Calhoun
USS Woodbury—#309—Commander Louis P. Davis
USS Nicholas—#311—Lieutenant Commander Herbert O. Roesch

DesDiv 31—ComDesDiv, Commander William S. Pye

USS Farragut—#300—DesDiv 31 Flagship—Lt. Commander John F. McClain

USS Fuller—#297—Lieutenant Commander Walter D. Seed
USS Percival—#298—Lieutenant Commander Calvin H. Cobb
USS Somers—#301—Commander William P. Gaddis
USS Chauncey— #296—Lieutenant Commander Richard H. Booth

DesDiv 32—ComDesDiv, Commander Walter G. Roper

USS Kennedy—#306—DesDiv 32 Flagship—Lt. Commander Robert E. Bell
USS Paul Hamilton—#307—Lieutenant Commander Tracy L. McCauley
USS Stoddert— #302—Lieutenant Commander Leslie E. Bratton
USS Thompson—#305—Lieutenant Commander Thomas A. Symington

The Reno Rescues the Cuba Survivors

Meanwhile, momentous events had been taking place farther south. It will be remembered that the Melville and several DDs got underway around midnight and stood out for San Diego. By the time Commodore Watson had gotten his speeding seahorses snugged down into a column formation for the night, Admiral Kittelle in his motherly Flagship was about to round Point Arguello with four cripples—the Selfridge, Farenholt (both from DesRon 12), Farquhar, and Wm. Jones trailing in her wake. The weather was foggy and she took soundings every 15 minutes. Another cripple, the J. F. Burnes, was somewhere back astern and was in sight of DesRon 12 which, as we know, had followed DesRon 11 out of the Golden Gate.

For the Reno was reserved an exciting bit of action which, as it developed, had a considerable bearing on the events of the night to come. Captain Dick Barry had put to sea from San Francisco to make his annual smoke prevention and full power trial en route to San Diego. Acting as umpire for this run was the Squadron Gunnery Officer, Lieutenant Commander H. K. Lewis, a tall, rangy Idaho lad whose poker face and slightly oriental eyes had won him the title of "Chink." He had been relieved of command of the Young by Commander Calhoun in August—a lucky shift for him.

After clearing the channel and passing the Lightship, the Reno pointed her prow southward and the black gang gradually warmed her turbines up to full speed. All went well and by 1330 she was about to head into Santa Barbara Channel. At that time she was about 75 miles ahead of the Melville.

"However," as Dick Barry said later, "the weather was getting pretty thick and I decided it was not good enough to risk the Channel making 30 knots because of shipping which I might encounter."

The International Rules of the Road require that, in reduced visibility a ship shall proceed at such a speed that she can come to a dead stop in half the distance that she can see. He therefore eased his rudder to starboard and swung the Reno's bow to seaward, intending to leave San Miguel Island—and the other Channel Islands—to port and continue on toward San Diego.

The Channel Islands, as they are locally known, form the southern boundary of Santa Barbara Channel, which runs some 63 miles along the southern California coast from Point Conception down to Point Hueneme. There are four of these islands— four rocky protrusions from the ocean's floor. To this day they are virtually as primitive and rugged as they were when Juan Cabrillo—who fell to his death on the rocks of San Miguel— discovered them in 1542. Their reef and rock infested shores have been the cause of numerous strandings. The many tales of sunken or buried riches still tempt treasure hunters and skindivers. San Miguel is the westernmost of this island chain.

Just as his ship was steadying on a new course for her home port, Fate took the helm in the interest of a lot of people whose lives otherwise might have been lost. The Reno's Quartermaster sighted a small boat, apparently full of people, and a few moments later, a second boat.

"I thought, at first," said Captain Barry in retelling the story, "that they were fishing boats, but when it was observed that one was waving a small white flag, I decided to investigate and headed back on the reverse course. We passed them close aboard, saw that they were in distress, stopped the full power run and lay to in their vicinity."

In his report of the rescue, Dick Barry wrote:

They proved to be two life boats from the S. S. Cuba, a Pacific Mail steamer out of Panama carrying passengers and cargo, which had grounded on the western end of San Miguel Island. [Author note: This vessel had grounded at 4:30 A.M. that day. What with her radio being out of order, she had not been able to ask for assistance. The First Mate was rowed in a lifeboat across the 20-mile-wide channel to Santa Barbara to give news of the disaster.] One boat was in charge of the Chief Engineer and the other in charge of the Second Officer. All of the survivors were taken on board, coffee and sandwiches served to them and every effort made to look out for their comfort.

The Chief Engineer stated that he did not know the exact position of the Cuba when she struck. He was of the opinion that he was heading across Santa Barbara Channel, in search of aid, at the time he was picked up. In fact, he was drifting in a southeasterly direction. He stated that most of the survivors had landed, he thought, on San Miguel Island and were in need of assistance as there were women in the party scantily clad. The Reno immediately started in search of the Cuba and when she was located off the western end of San Miguel Island the survivors could be seen huddled around a fire on the beach. The Reno was put in as close to the beach as it was thought safe, and anchored.

The motor sailer and whale boat were sent ashore, the former anchoring off a little way, while the latter transferred the people to her through a moderate surf. The spot in which it was necessary to anchor did not afford a particularly good lee, and there was a moderate sea running. Lieutenant (jg) G. A. Patterson was in charge of this operation and deserves great credit for the efficient manner in which the passengers were brought off from the beach under exceedingly difficult conditions.

The passengers, including 14 women, were gotten aboard and at 7:28 P.M., the Reno headed for San Diego. The wardroom and Chief Petty Officer's quarters were turned over to the women survivors and everything possible done for their comfort. Word was brought from the beach that the Captain and one member of the crew were still on board the Cuba, and about 15 members of the crew were left on shore as they did not desire to leave.

Word was also received that one boat in charge of the Chief Officer with about 15 men was still adrift. (This boat reached Santa Barbara.) It was at first intended to land the passengers at San Diego but the Chief Engineer of the Cuba stated that the

Pacific Mail Steam Ship Company had much better facilities for handling them in San Pedro, and as the Reno was off this harbor about 6:00 A.M., in a dense fog, it was decided to run in there and discharge the survivors. They were landed in a motor sailer from the Nevada at 7:00 A.M. Getting into San Pedro Harbor in pea soup fog was no cinch for the Reno. But she made it.

"Believe me," recalled Dick Barry recently, "it was not a pleasant experience getting into San Pedro. In that fog it was touch and go—much touch and very slow go—before we dropped anchor. When I at long last heard a cock crow, I was glad that I had memorized the standard order: 'Let go the starboard anchor.'

"When the fog cleared in the morning, I discovered to my great surprise that I was anchored smack in the middle of the Battle Fleet. The Lord certainly stood lookout watch aboard the Reno on that particular night."

The Reno had rescued 71 survivors. She proceeded to San Diego that Sunday forenoon. The Cuba, exposed as she was to the winds and seas of the open Pacific, was a total loss, as was much of her $400,000 cargo of coffee and silver bullion. Her stranding at this particular time created considerable confusion as regards the succeeding disastrous happenings of the night. Also it led to an argument which, had it not occurred, might conceivably have altered the whole course of events and prevented the writing of a black chapter in United States naval history.

The argument developed when Commander Roper, ComDesDiv 32—always a man of action, albeit one with many square corners to his disposition—upon interception of the Reno's dispatch regarding the wreck of the Cuba, called Commodore Watson by radio phone. He requested permission to take his entire division at full speed to her assistance. The Commodore evidently did not feel free, in the presence of his superior, Admiral Kittelle, to order such a movement, and was quite firm in his refusal to let ComDesDiv 32 take off. Commander Roper was considerably disgruntled thereby.

This had repercussions that, possibly, could be held responsible for the Tragedy at Honda. Admiral Kittelle, who was in the near vicinity, sent one of his cripples, the Selfridge, to aid the Cuba but, actually, there was nothing she could do for the unfortunate ship.

Navigation—By Guess and By God!

In the combination emergency cabin and chartroom of the Flagship Delphy, as she led her plunging consorts southward that evening, three men—Captain Watson, Lieutenant Commander Hunter, and Lieutenant (jg) Blodgett—pored over a chart. The clock on the bulkhead read 6:10. Chief Quartermaster Cummings stood behind his commanding officer with the Pacific Coast Light list in hand. The wind and sea were almost astern. The ship was yawing considerably. Fast action on the part of the steersman was necessary to keep her reasonably close to the course. Now and again, added vibration of the hull indicated speeding up of the propellers as a following sea lifted them closer to the surface.

Visibility was definitely worsening and, when a change of course to 150 degrees, true, had been signaled to the Squadron at 1430, Point Sur, only 4 to 5 miles distant according to dead reckoning, was not visible. Hence, a check on the actual speed of the Squadron, made good over the ground, could not be had. They must continue on, using dead reckoning positions until, by bearings from the RDF at Point Arguello or by soundings or actual sighting of the lighter hearing its diaphone foghorn—the Delphy's true position could be fixed. For entry, with her Squadron racing along at 20 knots, into the fog-ridden reaches of Santa Barbara Channel, such a definite fix was most important.

"Gentlemen," Commodore Watson was saying as he stepped the dividers along the charted course, "I feel that we have two factors in our favor: the wind and sea are pushing us along and, according to the Coast Pilot, we have a slight assist from the Japanese Current."

"Right, sir," agreed Lieutenant Commander Hunter, "that will take care of any loss of speed due to foul bottom or bad steering, or even some racing of the screws caused by the following sea."

"We should have a bearing soon," said Lieutenant (jg) Blodgett, "from NPK, the direction finder station at Point Arguello. Radio has been trying for some time to break in but Arguello has been pretty busy giving bearings to the Melville. She's probably close to Arguello. Been asking for cuts every few minutes. Fog must be bad down there."

Captain Hunter, an experienced navigator as well as an instructor in navigation at the Naval Academy, was not too happy about using radio bearings: "Subject to a lot of errors, personal and electronic," he said.

Lieutenant Blodgett was in no position to argue with a navigator of Hunter's experience, although he had served as assistant navigator on two tin-cans and had been a destroyer man from the very beginning of his naval career. Born in New England, where traditions of the sea are strong, Larry Blodgett had enlisted in the Naval Reserve in December 1917, the moment he arrived at the minimum age limit. His country was at war and the spirit of adventure plus his feelings of loyalty and patriotism swept him into the nearest recruiting office.

He rose rapidly in rating and, just after the Armistice, he was commissioned as an Ensign in the regular Navy. Blodgett was a lad who read and observed a lot and—as is the quality of youth—he was inclined to favor "new-fangled gadgets," such as the radio direction finder. Interrupting their deliberations came the shrill whistle of the voice tube from the radio room.

"NPK reports that we bore 320 true at 1815, sir," came the word.

"Aye, aye," replied Blodgett and, seizing his parallel rulers, laid down the bearing on the chart.

"Looks pretty good, Cap'n," he said, "pretty close to our DR position."

Watson and Hunter leaned close; checked the plot carefully.

"Hmmm," growled the latter, still taking a dim view.

"Hmmm," growled the Commodore. "Just the same, I wish we had a sonic depth finder, one of those new fathometers which the British have in their new destroyer leaders. Can't see why, in the more scientific type of gadgets, other nations always are ahead of us. Look at that echo-ranging 'Asdic' the Limeys developed during the War for locating enemy submarines..."

"Well, sir, research costs money and in peacetime, appropriations in our Navy have been mighty skimpy," shrugged Hunter.

"Hmmm," murmured Watson, moving toward the door. "Keep the bearings coming. Only about a couple of hours until we have to make that windy corner turn. No use to slow down to sound here—probably couldn't reach bottom in 300 fathoms..."

"No use at all, sir," agreed "Dolly" Hunter, "and we'd spoil the engineering run."

Watson nodded and was gone, leaving uneasy minds behind him. Blodgett's youthful countenance wore a worried look, as did the weather-lined, windburned face of Chief Quartermaster Cummings. Hunter, on the other hand, exuded confidence.

Throughout his entire career, "Dolly" Hunter had been a handy man with a sextant, a genius with a slide rule, and a master of mathematics. He could walk his dividers down a chart with the swift precision of a highly skilled surgeon making an incision. During World War I, Captain Hunter served, with distinction, as navigator aboard American destroyers based at Queenstown. For months on end, in all sorts of weather, mostly foul, he took his tin-can into Atlantic, North Sea, and Channel waters wherever her guns or depth-charges would do the most good. He built up a reputation of having the homing instinct of a river-bound salmon. One might say that the Delphy's skipper was the ideal type of destroyer commander. He was self-confident, decisive, and with just enough geniality to dull the cutting edge of his cockiness.

There is little doubt but that Captain Watson placed great reliance on Hunter's navigational judgment. And why not? He had a record which, until that September day, virtually bordered on magic infallibility, and he had just come from two years of teaching navigation at the Naval Academy.

As a person, "Dolly" Hunter was chubby, affable, and lighthearted. Despite his, perhaps, too strong self-assurance, he had a wide vein of tolerance in his make-up. He had a warm and ready smile. There was a quizzical gleam in his dark, deep-set eyes. His rather lofty brow was covered by a thick thatch of coal-black hair. Before the end of the year, the chemistry of worry was to turn it as white as a cresting breaker. From the very time Larry Blodgett reported for duty aboard the Delphy the friendship and respect between himself and his skipper began to blossom. Blodgett, too, was a "hound" for mathematics.

Before joining the Delphy in 1920, he had served as assistant navigator aboard the Belknap and the Case. In February he had been promoted from Ensign to Lieutenant (jg). Despite his admiration for his CO, young Blodgett had a mind of his own. Among matters on which he did not agree with his skipper was the latter's distrust of radio compass bearings. Surely, they were not absolutely reliable, he admitted. On the other hand, he thought, they were not strictly for the gulls. There were worried looks in other chartrooms of DesRon 11 that night.

The trouble which Admiral Kittelle's Flagship Melville seemed to behaving in finding the Channel entrance was not lost on the navigators. Likewise the Reno's report of the Cuba disaster had introduced a new factor which did not ordinarily present itself to captains of southbound vessels.

The distance from Point Conception to the reefs of San Miguel Island is only about 23 miles. A fast ship, badly out in her reckoning, might end her days on that grim pile of rocks. Many had read the Coast Pilot instructions which said: "In thick weather the greatest caution should be exercised and soundings should be taken frequently."

This made sense, because the California coast is very "steep-to"—the ocean floor rises sharply from 100 fathoms to no fathoms. A 5-mile set toward the beach, whether caused by current, bad steering, or otherwise, could mean the difference between safety and death.

"Well," observed one long-time tin-can skipper as he looked over his navigator's plotting, "I don't know what's in the Commodore's mind, but it looks to me as if he plans to run Santa Barbara Channel. Me, I like to do things the easy way; I'd head out past San Miguel into the clear. The Channel, with all its traffic and fishing boats, is sure no place for a speed run at night, and in a fog. Just where do you figure we are?"

Knowing his Captain's keen sense of humor, the navigator took a chance on an ancient wheeze. "Just about here, sir," he grinned, and placed his open palm on the chart.

"That, Mr. Blow," coldly remarked the skipper, "is not funny. Get me some bearings."

"Aye, aye, sir," replied the youngster, well knowing he would have a tough time carrying out that order.

Asking for radio bearings by anyone except the flagship was contrary to Squadron orders. Cluttering up the air with requests from a whole squadron could only result in confusion and possible dangerous errors. In addition, the radio operators on the DDs stood a "split phone" watch, guarding the battleship wave with one ear and the destroyer wave with the other. The RDF station used yet another wave length in its communications. Hence, asking for radio bearings not only violated Squadron orders but also forced the DD operator to unguard one wave length. Under such circumstances, the Squadron Flagship obviously assumed responsibility for the navigation. As in fact, she is required to do by Navy Regulations.

But how were the DD captains to carry out the provisions of those same Regulations which made them responsible for the navigation and safety of their own ships?

This problem was solved in several ships, as it should have been, by initiative on the part of commanding officers or navigators. One or two actually requested and received radio bearings; others unguarded the battleship wave, listened on NPK frequency, and intercepted bearings intended for the Delphy or other ships. The information which they received gave them small comfort.

The angle of incidence between the course of the Squadron and the radio bearings was small—never more than 10 degrees. Thus correct estimation of the advance made along the Squadron's track —allowing for a possible error of a degree or so in the RDF operator's reading—was anybody's guess. To quote a line often used in our own "little ships" as well as in those of our British cousins, the navigation was "by guess and by God."

Most navigators were worried because they believed their ships were not making 20 knots over the ground and therefore were not as far south as they should be. Others worried because their own compass courses indicated that the Delphy was steering to eastward of the course set—150 degrees, true. That could be very bad, indeed.

Finally, there was the reaction evidenced by Commander Roper, ComDesRon 32, in his flag boat Kennedy.

This DD had an exceptionally competent navigator, Lieutenant R. F. MacNally, who, risking censure, had instructed his "Sparks" to intercept all possible radio bearings sent out to the Squadron. The alert Sparks intercepted two sent to the Stoddert and three sent to the Delphy. All of them were between 320 and 333 degrees and indicated that the sending station was nearly dead ahead. When informed of this, Commander Roper felt that this was excellent navigation.

"He's heading right for Arguello," he remarked. "Intends to pick up the light beam or the foghorn, then sheer to westward and round the Point."

The tragic thing was that, when the remainder of the Squadron intercepted the Delphy's position report made to Admiral Kittelle at 2000—a report based on her dead reckoning, as became known later—not one Division Commander or ship's Captain challenged it.

All felt that the Delphy's information, because of her frequent radio bearings, must be better than theirs; that the Commodore knew what he was doing and, possibly, that their own positions were not well enough established to warrant a protest. This, of course, was the time to slow and take soundings. If Arguello were dead ahead, the ships were in 20 to 30 fathoms of water. If they were west of the Lighthouse, soundings of 50 to 60 fathoms would be obtained. A single protest might have resulted in such action.

None was made.

Verily the jinns of Honda plan with fiendish finesse.

An Error in Judgment

Again in the chartroom of the Delphy, three officers and a Chief Quartermaster looked over the chart, but now there was a sense of urgency—faces were drawn and anxious. The Chief Quartermaster made quick trips out onto the bridge to look for the beam of Arguello Light or to listen for the raucous blast of its diaphone. The clock on the bulkhead read 8:30. The night was dark and the haze in the air had reduced visibility to a little more than a mile. The dead reckoning position plotted on the chart for 2030—8:30 P.M.—showed the Delphy to be well to the south of Point Arguello. On the present course of the charging DesRon 11 the rocks of San Miguel and the stranded Cuba lay only short miles ahead, according to the Delphy's chart. A radio bearing from NPK was called up the tube: "330, true, sir."

Hunter swore under his breath. "Another impossible bearing, Commodore," he said, then called down to the radio room: "Tell them we are south of Arguello. Ask them for the reciprocal bearing!" After a wait of 10 minutes, his patience wore thin.

At 2035 the reciprocal came back—168—but even that passed 2 miles to eastward of their assumed position. Hunter's face was stormy. "That's a bit better," he said, "but they are still a couple of degrees off."

"The bearings have been erratic by a few degrees, Captain," said Blodgett, "but they have all put us north of the Point. How about slowing for a sounding, sir?"

The Captain shrugged and glanced at the Commodore. Captain Watson shook his head. Not much use, probably can't reach bottom, spoil our engineering run."

Two more bearings came in. They varied by 10 degrees and were to northward, but their reciprocals looked better. Captain Watson laid down the parallel ruler with which he had been experimenting on the chart.

"I think we are all right, Hunter. Maybe we will have better visibility and better luck in picking up Point Conception Light. At 2100 change course to ninety-five degrees, true."

"Aye, aye, sir."

The die was cast.

GERMAN BATTLESHIP SCHARNHORST

Photo # 24-P-90   Cdr. Edward H. Watson, USN, 1915

5 - HELLBENT FOR HONDA

DD #261-20:58 30 hours (Delphy)

THE RED SWEEP-HAND ON THE ILLUMINATED DIAL of the large clock on the forward bulkhead of the Delphy's bridge showed a span of exactly 90 seconds before the hour of 9 o'clock—the moment set by Captain Watson for the destroyers under his command to make the eastward swing from California's coastal waters toward the Santa Barbara Channel.

The compact glass- and sheet-steel-enclosed area of the Delphy's bridge lay in a twilight so low that it was prevented from being utterly black only by the ghostly green glow cast by a small light within the compass binnacle. As always, unless the occasion called for speech and action, the bridge was wrapped in absolute and motionless silence. An unfamiliar observer would have to look closely indeed to discover the seven men of the bridge watch who, immovable as statues, stood at their respective stations: The steersman at the helm, the quartermaster, the messenger, the signalman, and two lookouts, one in the port, the other in the starboard wing of the bridge.

Lastly, but not least in the order of his importance, there was the OOD, the Officer of the Deck, to whom is delegated the responsibilities of the Commanding Officer for the duration of his 4-hour watch.

Ensign John A. Morrow, less than 3 months out of the Naval Academy at Annapolis, had the Deck. To improve his outlook, Morrow stood at an open window directly forward of the steering wheel and engine-order telegraph. Night glasses hung by a leather strap around his neck as he, maintaining a deep-sea sailor's stance, rocked gently with the abrupt motions of the ship. His intense, keen young eyes roamed searchingly over the stretch of visibility that lay ahead.

From below came the rumble and vibration of a destroyer plowing through a wave-crested sea at 20 knots—a pace that covered a distance of 11 yards every time the bridge clock's red sweep-hand sped from second to second. From forward, through the wide-open bridge window, came the sharp and constant hiss as the Delphy's razor-like bow knifed through the rough water.

This was Ensign Morrow's first turn to stand watch as a fully qualified Officer of the Deck. He bore this great responsibility with a natural mixture of pride and concern. This was quite different from being Junior Officer of the Deck. As JOOD, he had been given only secondary responsibilities.

But now, as OOD, he was—for four solid hours—the eyes, ears, brains, and voice of a vessel that was not only a man-o'-war but also the Flagship of a Destroyer Squadron.

Well, thought Morrow, with grudging consolation, it was better to be in the van and head into the night at 20 knots than to keep station in the middle of a column of destroyers—only 147 yards apart from bow to fantail in terms of distance and a mere 13 seconds in terms of time.

On a night like this, in a sea like this, and aboard destroyers that yawed, rolled, and pitched as they made way at a speed equal to that of an auto doing 23 miles an hour, keeping station took skill as well as concentration.

As it was, most of the destroyers in the three divisions were not keeping taut station on this particular run. This laxity was due mainly to the fact that quite a few of the vessels had difficulty in maintaining a steady 20 knots on their cruising turbines. And, in view of engineering competition rules, it would be a black mark against a destroyer if her Engineer Officer had to resort to the use of his high-speed turbines in order to keep station.

Thus, under the cover of night, several vessels in the Squadron ran with distances of from 350 to 450 yards between ships. As things turned out, this was a blessing in disguise. In the chartroom, Captain Hunter and Lieutenant Blodgett made ready to step out upon the bridge. They had just checked their ship's chart for the new zero-nine-five course, which the Squadron Commander, Captain Watson, had ordered to be set at 2100 to make the approach to Santa Barbara Channel. As they emerged upon the bridge, Lieutenant Blodgett passed the change-of-course order to Ensign Morrow with instructions to signal the turn to the Squadron and to sound two blasts of the destroyer's deep-toned whistle as he put the rudder over.

When the red second-sweep joined the clock's minute hand at precisely 9 o'clock, the order was executed.

Given standard (15 degrees) left rudder, the speeding destroyer swung smartly to port in a sweeping curve. The churning swath of wake she left astern became a straight line as the jackstaff on the Delphy's bow pointed exactly 5 degrees south of east. That the wind was no longer nearly astern but on the gallant little four-stacker's port quarter, thus giving her additional roll and yaw, was a nuisance to be taken in her stride. In another 15 minutes or so, the Delphy—if all went well—would have covered the good half-dozen miles that should take her from the windswept open sea to the more protected waters of Santa Barbara Channel.

That is, if all went well!

If he had any doubts at all about the situation, "Dolly" Hunter probably felt that the 55-degree eastward turn had been made in plenty of time to avoid the dangers of San Miguel and other reef-studded islands along the southern edge of the Channel. He was confident that Point Arguello lay well to northward of him. As his vessel made the turn, Hunter crossed to the port wing of the bridge. He wanted to take a look at the line of destroyers coming up his wake and approaching the Delphy's well-defined turning point.

While the night was hazy, there was enough visibility for him to scan the broken row of running lights. He could, with a little difficulty, make the lights of the Percival. She was, he knew, the eighth ship in line. Allowing the standard distance of 250 yards from the foremast of one destroyer to the foremast of the next, this would mean a visibility of about a mile. The night was dark, the sky overcast, and the sea was moderate with whitecaps and heavy ground swells. Still, Hunter's practiced navigator's eye took note of the varying distances between the destroyers in line. The ships were not keeping good station—but, in view of conditions, this was not surprising. He estimated that there was at least a full mile between the Delphy and the Percival.

Based on that, there was little prospect of the Delphy sighting the beam of Point Conception Light to port as she entered the 20-mile width of the Santa Barbara Channel. But that was quite all right with "Dolly" Hunter. His chief regret was that no light at all was ever shown on San Miguel because there, in his opinion, the only real danger could lie in waiting. Ensign Morrow widened his stance to synchronize his sway with the new movements of the ship. Now that she was on an easterly heading, the Delphy rose and fell on serried ranks of spuming seas that poured across her low fantail or inundated her forecastle as she nosed into an extra big one. To the rumble of her turbines and the vibration of the hull was occasionally added the thunder of her propellers as a heavy ground swell lifted them close to the surface.

The red second sweep continued its measured merry-go-round on the face of the bulkhead clock. One—two—three minutes. And every time the sweep reached the 12 o'clock position, the Delphy had sped another 667 yards nearer to her tragic destiny. Four—five—minutes.

2105:00 HOURS

Without warning, the destroyer plunged into a heavy layer of fog. Morrow stepped forward and placed himself squarely in the open window to wring whatever information he could from zero-zero visibility.

Hunter, who, with Lieutenant Blodgett, was still on the bridge, arched an eyebrow and said: "Pea soup."

In a swift aside, it should be said here that, as a navigator famous for his homing instinct, "Dolly" Hunter held no fear of fog, be it thin as skimmed milk or thick as Devonshire cream.

Battleship sailors still remember how he, as navigator of the BB Idaho, nonchalantly took that mighty battlewagon into Anchorage, Alaska, despite a fog so thick that the seagulls were walking on it. As a navigator, Commander Hunter had the monolithic serenity of a lighthouse. Before Lieutenant Blodgett could answer his Captain's remark, both men heard a swift, grating sound. As if the Delphy's bottom were touching gravel. It was hardly more than a whisper. Then in lightning-quick succession came a series of violent bumps followed by the smashing crash of a head-on collision. Hunter, Blodgett, and the entire bridge personnel were flung against the forward bulkhead.

They landed on the deck in a tangled heap. Even before they gained their feet, Captain Hunter began to issue orders. The Helmsman was ordered to stop the engines. The Signalman was directed to turn on the breakdown lights. The Quartermaster leaped to the whistle pull and sounded the four-blast danger signal, while Ensign Morrow swung his weight onto the siren's control and sent out its piercing warning. Lieutenant Blodgett, who received a painfully injured left knee in the crash, dashed below and forward to determine the extent of the damage done to the destroyer's hull.

Meanwhile, bare seconds had elapsed since the moment of impact. As he scrambled to his feet, Hunter took a swift look about him. Forward and to port, there was no visibility. But on the starboard hand, through wisps of slowly churning fog, he glimpsed the massive outline of a towering rock.

There was, in his opinion, but one answer: DesRon 11 had come too far to the south before making the turn. The Delphy had hit San Miguel Island or one of its outlying reefs. Reaching the speaking tube to the radio room, Hunter ordered: "Signal to the Squadron and keep repeating it 'Keep clear to westward and Nine turn'!" In Navy code, the latter signal directed the destroyers to make simultaneously a 90-degree turn to the left.

In this manner, Hunter hoped to send the Delphy's sister ships northward into what he believed to be the safe depths of the Santa Barbara Channel.

Precious moments were slipping by and opportunities flew by the board beyond hope of recapture. Astern, the ships of the Squadron were catching up at the rate of 11 yards per second. Chief Signalman H. D. Cummings discovered that the Delphy's signal searchlight had been knocked out of commission by the force of the crash. To make matters worse, her main radio antenna had carried away. Fortunately, both of the destroyer's Chief Radiomen, L. B. Lattimore and C. B. Tipsworth, were in the radio shack (just below the bridge) at the time of the disaster. With cool heads and practiced fingers, they hastily rigged their main spark set to their radio-telephone antenna. Now Radioman William Murphy began sending.

However, the ship's continuous-wave set, which they were forced to use, had, unfortunately, a poor note. Its signals were difficult to read. Realizing this, Lattimore and Tipsworth shifted their spark set to the telephone antenna. Now Murphy sent the same signal by spark. But due to the variance in wave length caused by the smaller capacity of the telephone antenna his signals were rather faint. Still, they could be read. And Bill Murphy kept on pounding them out until he had to leave the radio shack on orders to abandon ship.

To supplement blinker and radio signals, hand torches were also used to send warnings to the other ships to stand off. In engine and boiler rooms, the impact that had tossed the bridge personnel into a heap hurled machinist's mates, water tenders, electricians, and firemen sprawling on the steel deck plates or into bulkheads and machinery.

The Chief Machinist's Mate, who stood watch in the forward engine room, quickly resumed his post before the instrument panel and its nearby bank of long-handled engine throttles. As his expert glance read the dials, the Chief discovered that the engines were speeding up and that the steam pressure was building. A split second before Captain Hunter rang the engines to a stop, the Chief closed the throttles and popped the safety valves.

Back on the bridge, almost breathless from sheer exertion, Lieutenant Blodgett reported that he had found the forehold dry up to and including the forward storeroom. Believing that the Delphy would float, Hunter ordered "All back, full" on the engine-order telegraph. In moments, the destroyer's turbines screamed at their wildest pitch; black smoke poured out of her stacks as flames roared under boilers to change water into superheated steam; the propellers churned the sea into foam but the Delphy did not move an inch. Lieutenant (jg) Richard H. Cruzen, the ship's Engineering Officer, had been talking in the galley passageway with Squadron Engineer "Blinky" Donald, when the crash came. Both were thrown to the deck.

When Cruzen scrambled to his feet he dashed for the forward engine-room where he found that the throttles were shut and that the engine-room was not taking water. However, on going into the after fireroom, he found the bilges flooding rapidly. The same was true in the forward fireroom. The situation was hopeless.

Since time immemorial, it has always been difficult for men to remain calmly in the bowels of a ship in deep distress. In the #1 fireroom, where he was on watch, D. F. T. Awtey, Water Tender 1st class, set a splendid example for the men in his gang.

When sea water began to spurt into the compartment, he quieted a budding panic. The result was that all hands remained at their respective stations, calm and confident, until word was passed to secure. This they did in a thorough and efficient manner. With water gaining ground below, Lieutenant Cruzen asked and obtained Captain Hunter's permission to turn the burners off. This was done and the personnel was ordered topside. Going on deck, he aided the Chief Water Tender in preventing a build-up of steam that might cause a boiler explosion by lifting the safety valves on all boilers.

This done, Lieutenant Cruzen returned to the forward engine room, where the Chief Machinist's Mate and his men had remained unflinchingly at their stations. This, despite the steady and increasingly rapid inrush of water in both compartments. He ordered the Chief and his men on deck. The Delphy's turbines had made their last run. Topside! A moderate degree of bedlam had been raised by the constant screams of the Delphy's siren and a string of strident blasts from her steam whistle. Then, as the generator ground to a stop, lights went out all over the vessel and the howling duet of siren and whistle died on fade-away notes. (1923 DDs did not have diesel or battery-powered emergency systems.)

From the port bridge wing, Chief Signalman Cummings saw the dim lights of the S. P. Lee grow brighter through the drifting vapors. The Division 33 Flagship was approaching the Squadron Flagship on what seemed to be a dead-sure collision course at a wide-open clip of 20 knots. Grabbing a megaphone and shoving the port lookout aside, Chief Cummings stuck his head and shoulders out through a port wing window and yelled at the top of his leather lungs: "Stop! Back up! We're on the rocks!"

But the S. P. Lee did not hear or need the warning. She had already taken full measures to avoid piling into the stricken flagship. Luckily, she missed the Delphy by a hair. At the time, the fog was too thick to allow an accurate evaluation of what had happened. But, for all practical purposes, the Delphy had crashed into a rugged wall of deeply serrated volcanic rock. She was not, as "Dolly" Hunter and others thought, stranded on or off San Miguel in the Santa Barbara Channel, but almost 40 miles north of that island.

Contrary to his calculations, Squadron 11 had been led deep into the Devil's Jaw just off the cliffs of Honda mesa. Instead of being some half-dozen miles off the entrance to the Channel at the time the course was changed, the Delphy had been a scant 2 miles off the California coast and above Point Arguello Light. And now the Delphy's mangled hull lay helpless in the drydock of the jinns of Honda against the northern wall of Bridge Rock, a small "island" as ugly and inhospitable as Honda itself.

The destroyer's bow was so badly telescoped that its deck plating had been thrust aft as far as the capstan. Also, her forward oil tanks were ripped open. The thick, viscous liquid floated in huge globs on both sides of the ship. Because of its density, the fuel oil did not spread rapidly.

In fact, the water between the Delphy's starboard side and the rock was soon covered by a blanket of oil so thick that it represented a major threat to the safety and survival of any human being who ventured into it. Almost immediately, under the hammering of the surf, the vessel was shoved to starboard and toward the almost invisible cliff. At the same moment, and with sardonic timing, a great sea lifted her as if she were a toy. It carried the destroyer over a submerged reef and dropped her after section upright in a cradle of rocks. The ship's bow and foredeck rested precariously on a rocky shelf. The only factor that saved her from immediate destruction, by sliding into deep water, was that her stern rested firmly in its cradle. Due to the combined action of a heavy surf and wind-driven seas, the stranded vessel was pounding so heavily that it was only a matter of time before the Delphy would break up. She might last a day or two, at best—but she would never sail again.

There was, Captain Hunter knew, nothing left to wait for; nothing left to hope for. He turned with a heavy heart toward Captain Watson, ComDesRon 11, who had come on the bridge from the chartroom right after his Flagship struck. Captain Watson, his face stony and his eyes as flat and expressionless as pieces of slate, regarded Hunter with a fixed stare. Then he nodded slowly. Neither man spoke a word. There was nothing to be said. The situation spoke for itself.

"Clear the bridge?" snapped Captain Hunter without a trace of his usually warm and friendly tone.

As the bridge personnel, followed by Captain Watson, scrambled down the ladder of the bridge structure to the deck, Hunter followed. Once on deck, he hurried aft to pass the word to "Abandon ship"—the two most dreaded and reluctantly spoken words in a sailor's vocabulary.

On the bridge, everything was dark and silent. Soon the bridge crew of Davy Jones would take over and guide the Delphy to her tomb in the Graveyard of Ships. The compass light had nickered out and the face of the bulkhead clock was no longer luminous. In the faint glow of a battery-lighted electric globe that had been left burning in the chartroom, one could see the face of the clock.

Its hour, minute, and second hands had run out of time for all eternity. But the ageless surf and the timeless seas smashed the Delphy back and forth—from starboard to port and from port to starboard; from starboard to port and back to starboard—with the careless violence of drunken giants rocking a cradle.

Throughout the ship, above the tumult of the sea and the hollow poundings of the hull, could be heard Captain Hunter's last order repeated like the playback of a multiple echo:

"Abandon ship! Abandon ship! All hands on deck! Abandon ship!"

Destroyer Division 33

DD #310—2100:13 hours (S. P. Lee)

The instant the S. P. Lee, flagship of Captain Robert Morris' Division 33, reached the turning point defined by the Delphy's turbulent wake, Lieutenant (jg) Arthur E. "Tony" Small, who had the deck, ordered the helm put over. Simultaneously, he signaled a left turn to the Young, the ship next in line, by two long-drawn blasts on the S. P. Lee's whistle. It was then 2100:13. This meant that the S. P. Lee, from the standpoint of station keeping, was right on the nose. Unfortunately, her Steersman of the Watch was a novice at the helm. By taking too wide a turn, he lost both distance and position.

Next, on finding himself too far to starboard on the new course, the man had overcorrected by giving the destroyer too much left rudder before he eased off the helm. The upshot was that the S. P. Lee was now on the port side of the Delphy and with a loss of about 200 yards.

In the 13-second interval between discovering the squadron leader's new course and reaching the turning point, Lieutenant Small sent his messenger on the double aft to the chartroom with information about the upcoming turn. This was not exactly news to Captain Robert Morris, Commander of Destroyer Division 33, or to Commander William H. Toaz, skipper of the S. P. Lee, which also carried the Division Commander's pennant.

They had heard, as had Lieutenant William E. Tarbutton, the S. P. Lee's Exec, and Navigator, the twin blasts of the Delphy whistle which signaled a left turn. Going out on the bridge they saw, when the S. P. Lee finally settled down on a new course in the wake of the leading destroyer, that the bearing was zero-nine-five. Captain Morris and Tarbutton re-entered the chartroom while Captain Toaz remained in the deep shadows of the after part of the bridge.

From where he stood, Bill Toaz was in direct line with the tiny navigation light that shone aft as a directional guide from the tip of the jackstaff. He heard Lieutenant Small, in the calm, impersonal voice of the OOD, say to the Steersman: "Right rudder, handsomely."

"Right rudder, handsomely," repeated the man at the wheel as he rolled the wheel a few spokes to starboard to make the correction.

Captain Toaz saw the jackstaff light swing gently to starboard, an act that brought the Delphy's fantail into direct line with the S. P. Lee's bow.

"Steady; steady as you go," came Small's clear, quiet voice.

As the Helmsman repeated the command, Small lifted the stadimeter and glued it to his right eye. The stadimeter is an optical instrument used quite constantly on the bridges of DDs and other men-o'-war in formation to determine their distances from each other. But "Toze" Toaz had long ago acquired the destroyer man's trick of figuring gaps between DDs by merely sighting over the tip of the jackstaff and drawing an invisible line between it and the ship ahead. If the line reached the fantail, his destroyer was in her proper station. If it went inboard, she was too close. If the line ended in the wake short of the fantail, the gap was too long.

A swift calculation made Captain Toaz place his ship some 400 yards astern the Delphy.

"What do you make it, Mr. Small?" inquired the skipper of his OOD.

"Sorry, sir! Exactly 450 yards. Or 200 yards short of station."

Captain Toaz uttered a disapproving "hmm."

He might have said more, a great deal more, but the full force of his attention was suddenly riveted upon the Delphy.

She seemed to have been swallowed by the night. The answer came moments later when the S. P. Lee entered the fog into which the Delphy had vanished. "Tony" Small stepped forward to gain a sharper view of the Delphy's already thinly spreading wake. Then, directly forward, out of the wool-gray fog and the murk of the night, came the lights of a ship.

First faintly; then sharper. It was as if the Delphy were backing toward them with her engines going full speed astern. She was, as Lieutenant Small saw it, literally leaping backward. Knowing that the situation was an optical illusion that could only have been caused by the Delphy's coming to a disastrously sudden full stop, such as running aground, Toaz at once ordered full speed astern and full left rudder. The voices of the Captain and his OOD spoke as one and in the same breath. In the flicker of an eyelash, the fog-smeared fore mast of the Delphy began to bloom like a Christmas tree as the breakdown lights went on and the steady white standard speed light atop her truck changed to a fixed red glow that commanded: "Stop!"

"Stop." That was easier to say than to do. At 11 yards per second—exactly the speed of a champion sprinter in running the 100-yard dash—the S. P. Lee's axe-sharp bow was cutting down distance, and at a breathtaking clip. Had the open-water gap between the two destroyers been the prescribed 350 yards of standard distance, a pile-up might not have been avoided.

To make things worse, the S. P. Lee's engine-order telegraph had barely ceased its repeat tinkling when the Delphy's lights were wiped out. There was nothing to indicate what had happened. Ahead reigned utter darkness. But out of it—as from a great distance—came the frantic wailing of a siren and the hoarse bellow of a whistle.

Then silence.

Meanwhile the S. P. Lee, in order to avoid smashing into her sister ship, slewed hard aport. She still held her forward momentum, but it was being reduced drastically by the reversed twin propellers braking her speed as they clawed to grip water with the fully mobilized might of their rapidly spinning 9-foot blades. Under the press of whirling turbines, and the violent resistance of tons upon tons of southbound seas that fought the destroyer's efforts to make a northerly turn, the S. P. Lee's light hull plating actually groaned under the terrific strain her rivets were taking. She skidded forward on her starboard bow while her stern barely missed the shapeless splotch of blackness that was the Delphy. When, an instant later, she came about on a nearly northerly heading, the S. P. Lee's speed had been reduced to about 8 knots. She had avoided the Delphy by nothing short of a miracle, a miracle, one might say, created by the inexpert steersmanship of a novice at the helm. To be sure, the S. P. Lee had missed more than the destroyer. But neither Lieutenant Small nor Commander Toaz nor Captain Morris was to know about that until quite a few more tension-charged moments had ticked into the deadwood of time. Hidden by the fog that hung like a fluttering curtain between the destroyer and visibility to the south, east, and north were the ragged outcroppings off Bridge Rock's northern shore, the nearly perpendicular cliffs of black volcanic rocks that rise to Honda mesa and the bulging bluff of Point Pedernales.

Had the S. P. Lee failed to arrest her eastward momentum, she would have crashed head-on into a massive, wave-pitted cliff—the forbidding face of Honda. As it was—with her whistle and siren going full blast to give their frantic warnings and her backing light flashing red—at 2106 the S. P. Lee struck. She had been backing for some 40 seconds, and the Captain estimated, from the shock, that her speed had been reduced to about 8 knots over the ground. As soon as it was evident that the DD would not back off, he stopped the engines and quickly went ahead slowly to try to swing her bow to seaward into the onrushing breakers.

That failed to work. In a few seconds, Captain Toaz felt a heavy shock. It telegraphed the bad news that the troubled vessel had struck again. This time farther aft and hard. He stopped the engines, asked for and received a report from the engine room. It was to the effect that the engines were functioning properly.

Skipper Toaz had barely passed this welcome information on to Captain Morris when Lieutenant (jg) Paul W. Steinhagen, his Engineer Officer, ran up on the bridge and asked permission to secure steaming boilers #3 and #4 in the after fireroom. Water was entering the room at a rapid rate. He added that the forward engine room was also taking water. However, the after engine and forward firerooms were still dry. This meant that, with luck, the #1 and #2 boilers could produce steam for the turbine in the after-engine room to get the S. P. Lee under way. By now the destroyer was receiving a veritable barrage on her port side from the ceaseless onslaught of the surf. Great seas broke constantly all the way across the deck. They also caused the ship to sway wildly from starboard to port—like a gigantic metronome—as it labored on sharp-toothed rocks that gnawed hungrily at her thin metal sides and bottom, flooding her fore and aft beyond hope of escape.

In the face of this display of restless force and uncontrollable motion, Lieutenant Steinhagen made his way aft from the bridge to the galley, climbed the galley housing, fought his way up the rungs of the #2 smoke-pipe where, with main strength and audacious courage, he removed the iron lid that covers a smokestack when it is not in use.

A few minutes later, Steinhagen was in the forward boiler room aiding Chief Water Tender O. H. O'Hara, L. M. Goodson and B. M. Barpart, Water Tenders 2nd class, and E. K. Patterson, Fireman 1st class, to run #2 boiler down to the steaming level.

This was accomplished. Steinhagen was just about to light off when Captain Toaz countermanded the order. No use. The after-engine room was in the hands of the enemy—the sea. Now the Devil's Jaw could really begin to chew its victim to pieces. As a good starter in that direction, a jagged tooth of volcanic rock crunched through the S. P. Lee's side and into the generator. As if by signal, all the vessel's lights went out. The S. P. Lee was not only rolling on the bottom but was also being worked bodily to starboard.

Once, when the curtain of fog lifted, those aboard the vessel could see a slim promise of survival only about 50 feet away. But to the grim-faced men, who hung on for dear life to anything stationary on their ship's steeply slanted deck, it looked as if the precipitous black wall rose to heights beyond their reach. The probing fingers of flashlights revealed that there was no beach or landing where wall and water met—only a narrow ledge of rock that ran along the precipice a foot or two above the level of the water. Between the ship and the shore, the sea ran so violently that it seemed impossible for either boat or swimmer to take a line across without being battered to bits upon the rocks. Captain Toaz knew that the time had come to abandon ship. Captain Morris agreed with him. But where to go? And how? They had no certain knowledge as to just where they were. But, at the moment, their best guess was that they were aground on or off one of the islands in the Santa Barbara Channel. Probably desolate San Miguel where the S. S. Cuba had been stranded that morning. This, for all practical purposes, placed them 1 mile short of nowhere. Or, to be more exact, just 50 feet west of oblivion.

DD #312—2055:00 hours (Young)

Aboard the Young everything had been made shipshape and Bristol fashion for the night's run which, God willing, would end at the San Diego Destroyer Base in the early hours of Sunday morning. After having completed and checked the Night Orders, Commander William L. Calhoun, the Young's brand-new skipper, and Lieutenant Eugene C. Herzinger, bis Exec, and Navigator, left the chartroom and went on the bridge to take a look at the situation. Ahead, the lights of the Delphy and the S. P. Lee were clearly visible. Astern, they checked off the running lights of five destroyers steaming in formation as straight as if they were drawn on a string.

If any one skipper among the destroyer commanders in DesRon 11 wanted—more than any other—to make a good showing on this competitive run, that skipper was probably Bill Calhoun. In 1923, he was in his late thirties, and a bit on the stout side. He had a rather ruddy complexion and his wavy blond hair was thinning on top. The Young was not only his first destroyer command but actually his first destroyer duty. He was a veteran battleship man and one of the first submarine skippers. Where some commanders might have tried to bluff their way through to conceal their ignorance of destroyers and their ways, Bill spoke freely about his lack of experience in "the boats."

He had the magnetic personality of a born leader and, in jig-time, Captain Calhoun had the full confidence of his officers and crew based on their respect for him, his judgment and ability.

To be sure, Bill knew how to make friends and make them accept his leadership without making a single concession to personal popularity. He was, on the contrary, a hard-driving go-getter. This was proven by his impressive collection of battleship gunnery and battle efficiency honors.

But, if the Navy had ever offered a trophy for durability in conversation, Captain Calhoun, who was a great-grandson of the marathon orator of the United States Senate John C. Calhoun, would have been its custodian. At first, the complement of the Young had a hard time making him out when Calhoun took command in July 1923. He was a pronounced contrast to Lieutenant Commander H. K. Lewis, whom he relieved, a man of very, very few words.

But it did not take Bill very long to win his way into the good opinion and high regard of all. Barely 2 months had elapsed since Commander Calhoun took command of the Young, and here he was on an important competitive run that demanded skill, resourcefulness, and initiative. Luckily for him, the Wardroom and Chief Petty Officers—as well as seasoned ratings—trusted him to the core. At the top of these stood Lieutenant Gene Herzinger, the Young's relentlessly efficient Exec, and the Captain's dependable right arm. After looking at the ink-black sea and the overcast sky for some 5 minutes, the two men parted. The hour was 2055.

Captain Calhoun headed for the wardroom and a cup of Enrique Torres' ironclad coffee. Lieutenant Herzinger returned to his chart with a cheery nod to Lieutenant August H. "Gus" Donaldson who had the duty as Officer of the Deck. Below, in engine and boiler rooms, men were standing watch and attending to the task of producing enough power to keep the Young rolling, as closely as she could, at a speed of 20 knots. Chief Electrician's Mate Kerrigan—the only electrician aboard (for the ship was running 25 per cent shorthanded)—checked with his good friend, Chief Machinist's Mate "Red" Hall, to see if anything was wanted before he turned in. No. Everything was routine.

"Well, keep her rolling, Red," grinned Kerrigan. "You wouldn't want anything to interfere with my getting into San Diego in the morning. I'm taking the wife to the hospital. That young Kerrigan is due to be launched 'most any time now."

"Don't you worry, George," chuckled "Red."

"You'll put in to the maternity ward in plenty of time as long as I'm running these sewing machines but, as a matter of record, I've heard that it isn't necessary for the pappy to be there for the launching. He just has to be there for the laying of the keel!"

Kerrigan headed for the bridge by way of the galley. Like most destroyer men, he was an inveterate coffee drinker. Leaning against the cook's counter he found "Pete," more formally listed on the Young's roster as Chief Boatswain's Mate Arthur Peterson. He was a man of medium build, well-muscled and tough as anchor cable. His long, lean face terminated in a jaw as big as a 10-pound rock and his unexpectedly hearty laugh rolled like Thor's thunder.

"George, how about a game of acey-deucy when you come below?" asked Pete, as his horny hand gave Kerrigan's back a well-meant but spine-cracking slap. "It'll help take your mind off that baby business."

"I'll give you a go at it, Pete," replied Kerrigan. "But I want to hit the bunk early. Tomorrow is going to be a busy day. First thing, I'm going to take my wife to the hospital. Next, the heir to the Kerrigan millions makes his entrance into the USA on California's Admission Day. Almost as good a day to be born as the Fourth of July."

Draining his coffee mug, the Chief Electrician made for the bridge. There he reported to Lieutenant Donaldson, who asked him to go forward and turn on the steering light on the jackstaff so as to make a mark for the steersman. The Chief Electrician went down on deck and headed for the forecastle. But as he went forward, he found that the Young was changing course.

With wind and wave on the port hand, as the destroyer took her new heading, Kerrigan saw towering seas climb to the deck and sweep the forecastle. He dashed back to the bridge and asked what had happened.

"We've changed course to east and are heading into Santa Barbara Channel," was Donaldson's reply.

"Good, sir," exclaimed Kerrigan with a broad grin. "I didn't dream we'd be that far down the coast by now. I'm sure glad that we got those new condenser tubes in Puget Sound from the Melville or we might not have been able to make San Diego by morning. The old girl is sure making knots."

"Yeah," chuckled Donaldson, "and don't forget the stork can make knots too!"

Chief Kerrigan's baby problem had, for days, been the main subject of conversation aboard the Young. There was betting, even money, that the youngster would arrive before, or after, the Young made port. You paid your money and you made your choice. Buttoning his foul-weather jacket and hitching up its collar, Kerrigan left, stuffed a fresh cud of chewing tobacco into his wide-lipped mouth and started forward, leaning against the wind as he went. The onslaught of the sea was so constant and so violent that George began to think that it would be better to head forward by going down through the wardroom country, rather than taking a chance on being given a ducking if he tried to go by way of the forecastle.

As he thought the matter over, chewing his cud of tobacco and steadying himself by leaning against the shield of the foredeck's 4-inch gun, he felt the Young rise under him.

It was a weird and frightening sensation—as if she were trying to surface like a submarine. Or sailing through the hard jolt of an earthquake. Kerrigan, belonging to a school of men who rattle out of balance just about as easily as icebergs, wondered what could be happening.

The destroyer rose out of the water, higher, higher, and higher. She staggered drunkenly for a split second and then fell back with the resounding slap of a belly-diving whale. Still in formation and with the lights of the S. P. Lee and the Delphy in plain sight as they steamed eastward, the Young had run over a submerged pinnacle reef. This deadly booby-trap inflicted deep gashes in the vessel's bottom on the starboard side as it pushed her forward and up. By virtue of her great momentum, the destroyer rode over the pinnacle without slackening speed.

On an occasion like that, a man usually thinks of doing what he has to do. Therefore, Kerrigan's thoughts instinctively turned toward the generator—his main responsibility—which provided all of the ship's auxiliary power and light. If that went on the fritz—then good-by all idea of reaching San Diego by daylight. With all possible speed, George worked his way toward the engine room hatch. He had just reached it when "Red" Hall jumped out of the opening with the swift energy of a jack-in-the-box.

"Gangway, Red," shouted George as he tried to brush Hall aside.

"You can't go down there," cried Hall. "There's fifteen feet of water in the engine room. Sure as hell, the ship is sinking."

When the Young reached and brushed past the submerged pinnacle reef, the S. P. Lee and the Delphy were still underway and doing 20 knots.

It was the Young's misfortune to hit the hidden fangs on the outer edge of the Devil's Jaw. Her hours of agony began at 2104, or about a minute before the Delphy struck. Just before the destroyer was ripped by this hidden reef, Captain Calhoun had gone down to the wardroom "head." As he returned to the wardroom about 2102, the OOD called down the voice pipe, "Captain, the Delphy has changed course to the left."

"How much?"

"About ninety degrees and without signal."

"Coming up," shouted Calhoun and dashed for his cabin to get his hat and a heavy coat. He was running for the ladder to the weather (or main) deck when he felt the ship surge upward and to port beneath his feet. His first thought was that the Young had been rammed. With an agility built up and maintained during years of negotiating ship's ladders—including those of the early submarines—he streaked up to the topside and gained the bridge as the main crash came on the starboard side.

"Who hit us, Gus?" he called to the OOD.

"No one, sir! But we hit something," was the steady reply. "Shall I back?"

"No," replied Calhoun.

A quick check showed Calhoun that his OOD had acted promptly and with excellent judgment. The rudder was right so as to take the dying Young out of the formation. The stop light was on, the whistle and the siren were being sounded. The breakdown lights flashed their warning. And then it was dark—dark as the inside of a rubber cow, as Navy balloonists of that day used to say.

The generator had been drowned out. Swift surveys and fast reporting revealed that deep gashes had been torn in the vessel's starboard side.

The Young was listing and filling rapidly. Still, there was always the hope that she would have enough undamaged watertight compartments to give her adequate buoyancy to remain afloat.

Around her, the folds of fog began to fall like a slow curtain on the last act of a drama. The Young, lashed by the seas and pushed by the wind, drifted helplessly with an increasingly steep list to starboard. Her wallowing progress, during which her bow was swung around, was halted when wind and wave lodged her against a submerged ledge.

"Sound general alarm," Calhoun directed Lieutenant Herzinger, "and pass the word to stand by to abandon ship."

The order was executed. But in the brief interval of 30 to 40 seconds, the Young's starboard list had increased from 30 to 45 degrees.

Reviewing the situation—with the speed of film frames flashing on a movie screen—Bill Calhoun realized that the men aboard the Young had no chance for boats and rafts or even life preservers. Their only hope was that the ship would not sink. The Captain pinned that faint concept of survival on the belief that the destroyer was resting on a submerged ledge.

Also, he had caught a glimpse of the Delphy going solidly aground on what appeared to be a huge rock. Since he believed that time and opportunity would be on the side of those who remained with the destroyer, orders to abandon ship were never put into execution.

As he wrote later, in his report on the grounding of the Young:

"I made the port side of the bridge by crawling and stood by the depth charge release gear. The bridge was absolutely quiet and no confusion existed. I saw no chance except to try to make the port side. I passed the word, repeated, to the Executive Officer and Chief Boatswain's Mate Peterson: "'Make for the port side!' 'Stick with the ship!' 'Do not jump!' "

Without batting an eye, slim, tough Gene Herzinger and equally tough Peterson opened a window on the uptilted side of the bridge wing.

First one, then the other, climbed through it, slid down the gently sloping side of the bridge structure, crawled up to and climbed the port rail.

On the starboard hand, they saw the eager seas reach up the Young's sharply listing deck with foaming fingers. On the other hand, was the slope of the destroyer's hull as it fell away from deckline toward her sharply molded bottom. Around them was the darkness of night; drifting banks of fog; the sharp sound of the cutting wind, like that of a scythe mowing wheat.

But within the two men burned the flames that give men the power to hope and the will to do, no matter how slim the odds in their favor. And the chances aboard the Young were, at that moment, slim indeed. Hanging on as best they could Gene and Pete crawled along the sloping port side of the ship, calling in steady, reassuring voices to the men who were clinging in small groups to the galley housing, the searchlight structure, and the after-gun platform. Many of them—most of them, in fact—were young and inexperienced. They believed the ship was doomed. So why not jump into the sea and swim for it? Where? Anywhere!

The human urge of action as opposed to inaction was exerting its hypnotic pull. But Herzinger and Peterson fought this impulse with all the power at their command. "Come on," they cajoled. "Obey the Captain's orders! Stick with the ship! Don't jump! Make for the port side! Lively now!"

The calm, yet urgent manner and voices of the two men won the day. At first, singly or in pairs, the men crawled up the port side and climbed the rail. Then the remainder joined in a rush. In seconds the deck was empty. But it was high time. As Captain Calhoun noted: "Very quickly—not over 1 minute and 30 seconds after she first struck—the Young heeled 90 degrees to starboard. She came to rest on her starboard side with her port side horizontal and only about 2 feet out of water. We seemed steady and fast on the rocks. Seas broke over us as we gathered the crew on the flat side of the hull just abaft the forecastle. I had the officers and chiefs quiet and encourage the crew while the Executive Officer, Chief Boatswain's Mate Peterson and I made plans for landing."

Bill Calhoun had placed his firm and wholehearted faith for the survival of the men of the Young on the existence of a God-given rock. And The Rock—a symbolic sign of rescue for men of the sea from the Ark of Noah down through the ages—would not fail them. The darkness was impenetrable. Even from the distance of a few feet, it was impossible to distinguish the crouching figures. The men were closely huddled in a small space for warmth, safety, and reassurance. Few of them were dressed. Many of them—tossed out of their bunks—were barefoot and clad in only skivvy shirts and shorts. Many had fallen overboard as they passed out of the hatches onto the almost vertical deck. All were soaked to the skin, and cold. By being closely massed, they hoped to prevent the everlasting onrush of surf and swells from sweeping them overboard. There was nothing to hang on to. And overhead was abysmal darkness descending upon them with the threat of a steel press.

Then, with the swiftness of a miracle, there was light. From somewhere—no one on the Young had any idea of its source—came a slanting beam of powerful light. Shouts, cries, and cheers arose from the men on the half-submerged Young. In stygian darkness, there is no greater harbinger of hope than light. And this shaft of brilliant illumination actually seemed to come from heaven.

As the light shone on the pitiful scene, the listing hull of a ship on which men waited with deep uncertainty for their ultimate fate, Peterson was struck with an idea that would improve their lot. His eyes caught the glint of glass in the 14 portholes that ran along the top edge of the port side between the Young's bow and foremast. Rising swiftly, he made for the port railing and slipped into the seas that surged over the destroyer's deck. As his wondering shipmates sat with bated breath, he vanished, only to return with a large fire-axe in his huge left hand. Brandishing the axe, Pete made for the portholes. One by one, with heavy, well-directed blows, he smashed the thick glass in them.

No one had to read a diagram to understand what it was all about. Creeping and crawling, all hands crossed the expanse of sloping steel to secure most necessary hand-holds. Inspired by his example and resourcefulness, the men stripped off the light hemp cordage which weaves a network between the life lines and lashed themselves in small groups to stanchions or anything that offered security. Men swarmed about each of the 14 ports of refuge like flies on lumps of sugar. They laughed. They shouted. They showed the sailor's traditional disregard of death, danger, and discomfort. One man, who had a lusty basso, sang: "Yes, we have no bananas!" Others changed the words in joining: "Oh, yes, we have no destroyers; we have no destroyers today!"

DD #309—2101:26 hours (Woodbury)

Commander Louis Poisson Davis, skipper of the Woodbury, wore many hats on that September day as DesRon 11 made its 20-knot run toward San Diego. Like most of the tin-cans in the Navy's Destroyers, Battle Fleet, the Woodbury was not only undermanned with respect to her enlisted personnel, but, instead of her full complement of seven wardroom officers, she had only five, plus the Division Medical Officer. And, among these, she did not have a single line officer above the grade of Ensign. Four of these were of the Naval Academy's 1921 vintage. The fifth had come out of Annapolis that very spring.

They were grand youngsters—charged with fizz and ginger— long on enthusiasm but short on experience. There were times when "Louie" felt very much like a mother duck with a newly hatched brood of chicks. As for the Division Medical Officer, like other Squadron or Division staff members, he rode wherever he could find a berth. The Flagships were all ordinary destroyers and badly rammed for space. During the gunnery exercises that morning, Captain Davis had doubled in brass as gunnery officer.

And, when the 20-knot run had got under way, he had lent a helping hand to Ensign Charles R. Pratt, his Exec, and Engineering Officer. In addition, he was his own Navigator. He jolly well had to be. It had been a busy, exacting, but far from unsatisfactory day for the CO of the Woodbury.

In engine and boiler rooms, all hands had shown enthusiastic interest in the competitive run. Steam pressures and speeds had been maintained despite occasional difficulties. The bridge personnel, too, had done a creditable job. The OODs and the steersmen, in particular, had been alert in keeping station and had held the course notwithstanding the destroyer's tendency to yaw with the following wind and the seas pushing hard from astern.

All in all, Captain Davis had good reason to reflect proudly on the fine teamwork of his young men as he shrugged into his foul-weather coat in his cabin abaft of and connected with the wardroom. The run was not over by any means. Still, if things kept going as they were, the Woodbury would show a creditable record at the end of the trial. But he was taking no chances.

At this very moment he was preparing to go top-side and spend the night on the bridge, in the engine and boiler rooms, if need be—or in the Captain's emergency cabin just abaft the bridge. He wanted to be instantly available, just in case. Through the open door that led from his cabin into the wardroom, Davis heard the peanut whistle of the speaking tube to the bridge.

"Captain speaking!" he said, answering its summons.

"Sir," came the voice of Ensign Bushnell, who had the deck, "the Delphy has just made a sharp turn to port. As far as I can judge, the new course is about due east."

"Coming right up," replied Davis.

Stepping quickly, he returned to his cabin for his cap, gulped down the now lukewarm dregs of what had once been a mugful of steaming coffee, and headed topside on the double.

When Captain Davis stepped upon his bridge, the Woodbury had already made her turn. He did not know what to expect, but, as a navigator thoroughly familiar with the California coast, he knew that they must be rounding Point Arguello Light. This beam sweeps the sea for miles around from a point some 2 miles south of Honda. It was not in sight.

"Run up to the crow's-nest," he ordered the lookout in the starboard bridge wing. "Keep your eyes peeled to port for a light-house beam. Report its position the moment you lay eyes on it."

With a snappy "Aye, aye, sir!" the man made a dash for the ladder leading up the foremast which stood directly abaft the bridge.

Simultaneously, Captain Davis ordered the port lookout to keep a bright lookout on the port bow for a flashing white light.

The destroyer skipper had barely given this order when Ensign Bushnell reported, as calmly as he could: "Sir, I have lost sight of the Delphy. The S. P. Lee has sheered out to port and the Young has stopped."

Louie Davis literally covered the distance to the forward windows of the bridge in one jump. Directly ahead, he saw the Young apparently motionless. To the left he caught a glimpse of the S. P.

Lee. The Division Flagship's truck still showed the steady white light that indicated standard speed.

Then the S. P. Lee, like the Delphy, was unaccountably whisked out of sight. Ahead lay the Young. One moment her lights were shining; the next she was dark.

Meanwhile, the Woodbury was bearing down upon her at a speed of 20 knots. Eleven yards per second and little more than 100 yards to go. Captain Davis ordered full right rudder to avoid the Young and to keep clear of the S. P. Lee. The speedy little sea-hornet heeled hard to starboard, cutting a wide swath of boiling wake as she started to make a tight right turn. By the time the vessel's head had swung about 10 degrees, the Woodbury struck and apparently jumped over a rock.

"I reached for the engine telegraph to back full speed," said Captain Davis, "but, by the time I had them on 'Stop,' the ship crashed hard on the rocks with the starboard bow against a rocky islet. I ordered the whistle and siren sounded as a warning to ships astern and ran out onto the starboard wing of the bridge to get a better look."

Fog was closing in around the ship, but with the aid of the main searchlight, Captain Davis discovered, with a sinking sensation in the pit of his stomach, exactly what he was up against.

To the right of the destroyer's broken jackstaff was a huge boulder. It is formed of the same igneous rock that the teeth in the Devil's Jaw are made of. Davis noted that the small island, if it could be honored with such a name, was very precipitous, jagged, full of deep depressions, and guarded by several detached rocks and ledges.

All in all, the spot which from that day on became known as Woodbury Rock was about as villainous a reef as ever sent a sailor's heart into his throat. To port, the destroyer leaned against a snag-toothed pinnacle rock. Captain Davis believed that his vessel had struck so hard and appeared to be so wedged upon the rocks that she could not be gotten off under her own power. Still, he determined that nothing could be lost by making an attempt to work free. However, being a skipper who knew the value of the lives entrusted to his care, he decided to play safe and to use Woodbury Rock as an anchor to leeward in case the destroyer should get into a worse situation.

It was possible that he might be compelled to abandon ship. In that case, Woodbury Rock would be a handy, if not a pleasant, refuge. Since, at the moment, there was a gap of only from 3 to 6 feet between the destroyer's slowly swinging bow and the nearest wall of rock, Davis decided to take advantage of the opportunity by putting men ashore to handle life lines if they should be needed.

With his megaphone, Commander Davis called down to the throng of off-duty men who had poured up from below following the crash. He explained the situation and called for volunteers. Instantly, a number of men pushed forward. Among them was Chief Boatswain's Mate Paul E. Pointer, a rugged old timer with hashmarks halfway up his sleeve. Addressing his words to Pointer, the Captain continued:

"Take four men with you on the rock and rig lines to steady the bow. We may have to use them to abandon ship. Be careful; put life jackets and lines on the men and don't let anyone get caught between the ship and the rocks. We don't want any casualties!"

Pointer promptly selected Commissary Steward Matthew H. Ryan together with Silas A. Puddy, Ralph C. Kiblinger, and Bob M. Wolf, all Seamen 2nd class.

Equipped with electric torches, the quintet ran forward and, one by one, made the leap from the bow of the destroyer to the cold, slippery, wave-swept inferno of Woodbury Rock. In the hard glare of the searchlight, which had been manned by Seaman 2nd class Evans W. Watkins, the rock had the unearthly look of a miniature satellite in space. It was a scene of sharp contrasts— like noontime on the moon. Tall or stubby spines and spires of volcanic rock, each with edges as sharp as a dagger, stood bathed in bright light and cast coal-black shadows. Between these monstrosities ran low ridges and deep chasms. The latter were pitfalls where men could easily plunge to maiming or death.

Once ashore, Chief Pointer signaled Watkins to take his searchlight off. He and his men would get along better from then on with the torches they had brought along. Slowly, Watkins worked his searchlight's beam to starboard. There was nothing there. Then, he swung it to port only to discover that his efforts to probe into the night with his blade of light were frustrated by a swirling mixture of fog and smoke. He ran the light back and forth along the narrow horizon over the port bow. Suddenly he saw something on the surface of the sea that was darker than the night. He sighted his beam upon it.

It was the Young. She was listing so far to starboard that only her port side was showing. A closer look revealed that the motionless mass that rested on her side was made up of men who were hanging on for dear life. The sight spread before Watkins' awe-struck eyes only a few minutes. Then layers of fog wiped out the gruesome spectacle as if it had only been a nightmare vision. Meanwhile, on the Woodbury's bridge, Ensign Horatio Ridout (Class of 1923) reported to Captain Davis that three forward compartments (A-1, A-2, and A-3) were completely flooded. Water was entering #1 boiler room and the forward engine room. But all compartments above the second platform forward and the first platform aft were dry. Davis ordered full speed astern.

Black smoke poured out of the destroyer's smokestacks and her propellers rotated at maximum revolutions as enginemen, firemen, and water tenders, under the direction of Ensign Ridout and Chief Machinist's Mate M. E. Tillett, labored with almost superhuman skill and courage to produce the steam and power to wrench their ship free from the grip of the rocks. In the face of deadly danger from exploding boilers, superheated steam and a sudden inrush of the sea, this score of men sweated at their tasks until 2230 hours, when all power failed due to flooding of the Woodbury's hull.

At that hour, Captain Davis decided to take action to start an evacuation from his ship that was to make heroic history on Woodbury Rock.

DD #311—2101:39 hours (Nicholas)

The Nicholas was steaming down the white phosphorescent carpet flung successively over the heaving surface of the inky sea by the speeding propellers of the Delphy, the S. P. Lee, the Young, and the Woodbury. Between her—the last of the four destroyers in Captain Morris' Division 33—and the Squadron Flagship was a distance of about 1400 yards and 99 seconds. In her steaming fireroom, the gauges on the destroyer's massive

Yarrow "Express" boilers, with a round dozen oil jets burning brightly under each, were registering the standard working pressure of 265 pounds.

The superheated steam flowed through 10½-inch steam pipes, through the throttles, and into the low-pressure high-speed Curtis cruising turbines in the two engine rooms abaft the boiler rooms. In the turbines, vaned wheels spun at the dizzy rate of 3000 revolutions per minute. From the time when the 20-knot run to San Diego began, Chief Machinist's Mate C. L. DeWitt and Chief Water Tender R. A. Moore kept their respective gangs on their toes to maintain a steady rate of 3000 rpm. This was the top speed the cruising turbines could deliver. It took a lot of babying, and expert anticipation, to maintain it. But the men in the bowels of the Nicholas had a great deal of pride in their ship and its equipment. A DD's power plant, the men swore, just was not meant to be mishandled that way during piping times of peace. If they, meaning the Brass Hats, had a yen to make knots—so groused the men who were exposed to the full concert pitch of the earsplitting cacophony of turbines, pumps, and condensers; why in hell didn't they put on the main turbines, jump the revs to 3200 per minute, and bring the props to their maximum of 452 RPM? At that rate, the Nicholas and other DDs of her class were supposed to reach a high speed of 34 knots. Well, maybe not. But, anyway, she'd skim the sea at a good 32 knots. The men in the engine and fire rooms had no way of knowing that the Nicholas was approaching a change of course at 2101:39 hours.

Their first news of it came when the ship heeled to port as she swung on her pivoting point, remained sharply canted for a dozen seconds, and then straightened as she stood on her new course.

"Feels to me, sir," yelled Chief DeWitt to Ensign Rae Cunningham, Assistant Engineering Officer, who had the watch, "like we made a turn of about forty-five degrees."

"Just about that, Chief," agreed Cunningham in his slow-spoken North Carolina accent. Being a very junior Ensign (Class 1923), his sea legs were not yet so finely attuned to the feel of the ship that they could convey such information to his head in a matter of seconds. In the wink of an eye, Cunningham had all he could do to maintain his stance before the instrument panel. Up to now, it seemed to him, the Nicholas had been leaping over the spume crested following seas much in the manner of a mountain sheep jumping from crag to crag. These violent motions were interspersed with frequent deep dips into the trough of the sea and steep climbs toward the heavens. It had taken Cunningham all the powers of contortion at his command to maintain his balance. But now that the destroyer had veered eastward and ran with the seas on the port quarter she leaped and lunged like a cat on a hot griddle.

"This can't last long," grinned DeWitt, who had troubles with his own footwork. "We must be heading into Santa Barbara Channel, sir. And then we'll be out of this."

Up on the bridge of the Nicholas, Charlie Reed, Torpedoman 2nd class, had the wheel. He was a bright, energetic youngster who took great pride in his appearance, his tin-can, and his steersmanship. No OOD had to tell him to "mind his rudder."

Due to the tendencies of a quartering sea to cause the destroyer to yaw, Steersman Reed had to use quite a bit of rudder from time to time to keep her compass course of zero-nine-five.

Lieutenant Commander Herbert O. "Fats" Roesch, skipper of the Nicholas, had come on the bridge just before the destroyer reached her turn, in answer to a call given him by Lieutenant H. F. Sasse, Acting Executive Officer. The ship was right on the spot with respect to keeping her station. Just 250 yards beyond, from truck to truck, was the Woodbury, in plain sight despite the growing haze. Ahead, but not so plainly seen, were the Young and then the S. P. Lee. The Delphy, oddly enough, was not in view. Suddenly, the firm picture of three destroyers steaming in column through the night disintegrated before Roesch's unbelieving eyes.

The lights of the Woodbury streaked to the right; those of the Young grew brighter; a sign that she must have slowed or come to a stop. Far ahead, the S. P. Lee's lights were turning left into the wind. All of these actions were signs of dangerous confusion. They were given additional emphasis by thick clouds of smoke and torching flames that belched from the stacks of backing ships, drifting curtains of fog that seemed to respond to invisible pulleys, and the somewhat muted chorus of screaming sirens and tooting whistles.

To avoid collision with the Woodbury and the Young, Captain Roesch reacted instantly with an order to Steersman Reed to give the ship left full rudder. As Reed executed the order, the destroyer heeled hard to port. Her bow hung in momentary suspense as the screws fought to help the rudder make the turn against the resisting forces of wind and sea. The battle was won by engine power in the ticking of a few long, long seconds.

"Now, what the hell do you make of that?" exclaimed Chief DeWitt to William Dearbane, Engineman 1st class and a veritable wizard when it came to cajoling turbines into doing their stuff.

"Somebody's gone nuts up topside," agreed Dearbane as both hung on to the railing of the low-pressure turbine in the #2 engine room. Even for seasoned destroyer men, the going was mighty rough as the Nicholas strove with all her might to claw her way northward and into the wind away from the hidden dangers of the night. The vessel plunged and reared like a bucking horse at a championship rodeo.

"Wonder what's cooking," answered the Chief thoughtfully. He added: "By the feel of the turn, she's heading north. Probably goodbye to our twenty-knot run."

"And good riddance," intoned Dearbane. "I've had my fill of this for today."

Just then the Nicholas crossed the submerged northern rim of the Devil's Jaw; about at the point where a cuspid should he in waiting. And it was there. The rock struck aft at the destroyer's starboard propeller. Instantly, the racing of the starboard shaft spelled bad news and the throttle was closed. At that moment, the pointer on the engine order telegraph raced to "Stop."

"Now, what in the devil's name was that!" DeWitt's voice rose to a full-throated roar.

Before the Engineman could answer, Ensign Nathan Green, the ship's Engineering Officer, slid down the engine room ladder with the speed of a bear on a greased pole. Despite his less than average size, he moved with hurricane speed as he ran across the steel floor plates, banged Ensign Cunningham on the shoulder, and with his usual Tennessee drawl forgotten, shouted: "I relieve you! Run aft to the fantail and steering gear room; find out if any damage was done to the propellers, rudder or steering gear. And report to me on the double. DeWitt, have a look in the shaft tunnels!"

The North Carolinian made off up the ladder while enginemen dashed back into the shaft alleys. Quickly, from the after-engine room came the bad news that the starboard propeller shaft was parted at the coupling in the shaft tunnel and that the shaft had been hauled aft about a foot and a half.

The rudder and port propeller seemed undamaged, reported Ensign Cunningham down the hatch. Green passed the information to the bridge. The starboard propeller was as useless as a broken eggbeater. The engine-order telegraph rang. "Back two-thirds," read the indicator. The appropriate levers were snapped into position and the Nicholas moved slowly astern.

Between stopping engines and reversing her direction, the destroyer had lost steerage way. Her bow, responding to the pressures of the elements, swung slowly from north toward west in the direction from which she had been coming. Again she struck rocks.

These were barely hidden by the surface of the sea, for the time of low tide was at hand. The poor Nicholas never had a chance. She drifted slowly astern and came to a stop, stern high, on a clump of rocks. With the ocean welling in from rents in her sides and bottom, the engine room and fireroom crews remained stoically at their tasks to carry out the orders of the steadily tinkling engine telegraph. Captain Roesch did his excellent best, but the jinns of Honda were against him. With all power available on his one effective propeller, he went ahead and he backed with full left rudder, full right, and rudder amidships. He strove in every way possible to get his ship clear, but without effect. At 2121, Ensign Green reported that engine and firerooms had been secured; fires out, steam pressure blown down, and safety valves opened. When Chief DeWitt and Chief Moore led their bone-weary but steadfast engine and boiler gangs up into the welcome chill of the night, sea water covered the floor-plates in both the engine rooms as well as the steaming boiler room.

Just as they came on deck, and slumped in the lee of the galley housing on the starboard side for a few deep pulls of fresh air, here was the hiss and roar as twin rockets soared aloft from the foredeck. When they exploded high above the Nicholas, their intensely bright flares illuminated the scene with terrifying effect. Large and small pinnacle rocks, laid bare by the outgoing tide, were sticking out of the sea on either hand.

About 100 yards astern showed the port side of the stranded S. P. Lee. That much did the fog let the flares of the rockets reveal, and no more. Due to the position of the Nicholas and the heavy breakers that roared down on her, the ship's company was ordered to keep to the ship and rig life lines. What with her 30-degree list to starboard, all hands were ordered to hold on to the port side in case the ship turned over.

"As it was not intended to abandon ship, unless compelled to during the night," wrote Captain Roesch in his report, "all hands were ordered to make themselves as comfortable as possible. No boats were lowered."

Being, as she was, on the windward side of Point Pedernales about 100 yards from Honda's hostile shore, the Nicholas was the target of the constant and fierce assaults of the thundering surf and the surging wind.

Great columns of water shot up her slanting port side, towered over her listing deck, broke and cascaded tons of deep-green water on the hapless destroyer. Soon, the drenched and shivering members of the engine room and fireroom gangs—seeking such poor shelter with their shipmates as the exposed deck afforded—wished with heartfelt fervor that they could return to the hot and stuffy comforts of the now flooded steel caverns in the bowels of their battered ship.

Destroyer Division 31

DD #300—2101:52 hours (Farragut)

The Farragut—leading destroyer of Division 31 and Flagship of Commander William S. Pye, Division Commander—steamed down course one-five-zero with her brood, the Fuller, Percival, Somers, and Chauncey coming up astern. Chunky, cordial, and capable Lieutenant Commander John F. McClain, the Farragut's skipper, known to his contemporaries as "Briggs," had come up on the bridge from the wardroom about 2000 hours for one of his frequent look-sees.

After a brief glance up and down the line of yawing destroyers, he turned to Lieutenant (jg) Charles C. "Chick" Hartman, who had the deck.

"Keep a sharp lookout to port for Point Arguello Light," he said. "The minute you make it out, let me know. I'll be in here," the Captain concluded as he opened the door to the chartroom abaft the bridge. Lieutenant Hartman acknowledged the order and alerted his lookouts. A matter of seconds past 2100, Lieutenant Hartman sighted—and his lookouts confirmed—what seemed to be a flashing white light on the port bow. The gleam was somewhat feeble, but the visibility had grown poorer during the past hour.

While the flicker in the night had none of the characteristics of a lighthouse beam, young Hartman sent his messenger on the double with word to the skipper. McClain hurried out on the bridge and gazed steadily in the direction designated by his OOD. By now, the pattern was changing. He saw what at first appeared to be a slowly lengthening line of shore lights. But that could not be possible.

Then he realized that the Delphy, spearhead of the squadron column, had made a sharp left turn and that the extending line of lights was formed by destroyers as they conformed to her new course. As Division 31's Flagship approached the point where she would follow the wake of the Nicholas around the turn, Captain McClain reported the change of course to the Division Commander who came up from his cabin in the bridge structure and joined McClain just as two long blasts of the Farragut's whistle signaled the turn to the Fuller some 400 yards astern.

As the Farragut settled on her new heading, Commander Pye and Lieutenant Commander McClain made for the chartroom to consider the change and to determine where zero-nine-five would be taking them. They were just bending to that task when the Quartermaster of the Watch reported sighting searchlights, hearing sirens, and the confusion of ships just ahead.

"Lieutenant Hartman," volunteered the Quartermaster, "believes it is a collision, sir!" Head down, running at full tilt, as if he were a halfback, "Briggs" dove for the open window in the forward portion of the destroyer's bridge. The OOD stepped nimbly aside to avoid his skipper's headlong rush. Coming to a quick stop, the Captain stared with outthrust head at the spectacle before him.

As he explained it later, "I saw ships steering out of column in numerous directions. I went ahead standard speed about ten seconds longer when I saw that we were closing up rapidly."

During those 10 seconds, the Farragut ate up 110 yards of the 150 yards of open water between her bow and the fantail of the Nicholas.

Captain McClain ordered speed reduced to two-thirds. Realizing that he was still gaining alarmingly fast, he rang for "Stop."

Even as the engine-order telegraph acknowledged the change, he called for engines "Full astern" and ordered "Hard right rudder."

Lieutenant Hartman took fast action to have the light atop the truck show a red reversing signal. The Farragut lost forward momentum as her screws bit into the cold green seas and churned them into a vortex of swirling foam. When his vessel came dead in the water, McClain stopped the engines and made a quick survey of the situation. About 100 yards ahead and some 30 degrees on his starboard bow, he saw a destroyer. She appeared to be sinking, and very rapidly.

Under the impression that she had been rammed—the destroyer was the Young just as she was starting to turn turtle—McClain called away boats to go to the rescue. Coming faintly through the night, he could hear the cries of men calling for help. As these screams rang in his ears and tore at his heart, McClain thought he caught a glimpse of a whaleboat some 50 yards just forward of the starboard beam.

"Boat ahoy!" he called through cupped hands. "Come alongside!" There was no answer. No dipping of oars. The boat did not move.

Then, from one of the madly reversing destroyers, trying so desperately to free themselves from the Devil's Jaw, came the quick, explosive glow of a torching stack. The flame, arising out of a smokestack, had the swift illuminating effect of a crimson flash bulb. By its ruddy glare, McClain saw that what he had taken to be a whaleboat full of men was actually a large rock pregnant with potential disaster. Simultaneously, he realized that every destroyer within sight was stranded. Aghast, he shouted: "My God! They're all aground!"

At that instant, it seemed to "Briggs" McClain as if he were looking at a terrifying seascape of painted wrecks upon a painted ocean.

The sight had a nightmare pattern that lacked reality. But when he again heard the yells of men calling for help at the top of their lungs, "Briggs" was brought back to grim acceptance of the cruel facts.

And foremost among these facts were that the Farragut, dead in the water, was drifting closer to destruction and that a Captain's first responsibility is for the safety of his ship and the men under his command. All other considerations are secondary. Much as he wanted to rescue the unfortunates who crouched upon the hull of the Young or floundered upon the oil-smeared seas, McClain knew that his ship was in dangerous waters. He had no choice other than to get out of them as quickly as he could.

"All back emergency," he ordered. So much zeal did the men in the engine room put into the execution of this order that their throwing of the throttles wide open robbed the generator of steam and it momentarily died.

All lights went out. Even the backing lights blacked out. Again Fate dealt from the bottom of the deck. The Fuller was coming up, now less than 200 yards astern. As the Farragut continued to back, McClain expected the Fuller to take the same action. But, as just mentioned, no backing light showed on the blacked-out Flagship's truck. Even if it had, the light would have had no effect on the Fullers behavior. Just brief moments before, while backing one-third to slow down her 20-knot rate of advance, she had struck a submerged rock which cut deep gashes in her sides and bottom.

Water had poured into her steaming fireroom and the forward engine room bilge so rapidly, that before the Chief Machinist's Mate on watch could inform the bridge, the ship had lost all power.

This happened only moments after the Fullers Captain had ordered "Emergency astern." Fortunately, she had already lost most of her headway. But she was still dashing forward at a fairly high rate of speed. Commander Pye, who had been keeping his seaman's eye astern, saw the onrushing danger. He warned "Briggs" McClain, who was concentrating on getting his vessel out of a threatening position.

With the Fuller bearing down upon the blacked-out Farragut's fantail, the skipper sounded his siren and put his destroyer two-thirds speed ahead on his starboard engine to prevent serious damage to either ship by a collision that seemed inevitable. As McClain threw the Division Flagship's stern around, the Fuller came within the thickness of a sailmaker's needle of her propeller guards.

Before anyone could bat an eye, the Fuller's bow sliced toward the other vessel's starboard side. As bad luck, in an overpowering deluge of misfortune, would arrange it, the Fullers bow hit the Farragut amidships. The latter's whaleboat took the real sting out of the glancing blow. It was damaged beyond use. The runaway Fuller dashed off leaving the damaged whaleboat and a deep dent in the Farragut's side, and carrying away life rails, awning stanchions, and stretchers.

McClain sounded "Collision quarters" and "Abandon ship," mainly to get all hands on deck—just in case. With the strain off the main engines, the generator came back to life and the lights functioned once more. "Briggs," with wonderful judgment and speed, figured a way to get out of his desperate situation.

Stranded destroyers ahead. Destroyers coming up from astern. And disastrous evidence of ship-killing rocks on every hand. He decided that if he could back out, on a course approximately opposite the one he came in on, he had a good chance of bringing the Farragut out in the clear. So he backed full speed westward on a course of about two-eight-five, true. But the Farragut was not to work her way out of the Devil's Jaw without being nipped as she ran. She was making good sternway when a shudder ran through the vessel.

She was lifted bodily by a great wave and dropped, with a resounding thump, on the rock-strewn bottom. Throughout the ship, all hearts stood still. Had she been badly mauled? Would she be able to float?

Or was she about to join those of her sister-ships who lay crippled on unidentified rocks? But, bless her stout heart, the Farragut kept moving astern as her propellers pulled at top speed. Swift examinations by damage-control parties revealed that no serious injuries had been inflicted.

Meanwhile, men with the hand leads were taking continual soundings. When the call rang out–"By the deep sixteen!"–Commander Pye directed McClain to come to anchor. At that time, the swirling clouds of smoke and fog lifted long enough to allow McClain to catch an approximate bearing on Point Arguello Light—120.

With the aid of his searchlight, he also glimpsed a large rock about 700 yards on the port bow with a stranded four-stacker just beyond it.

It had been McClain's intention to send boats to the stranded vessels. But it was too risky. Also, the Farragut's position was not good enough. McClain tried to slip the anchor chain and get farther out to sea, but the chain fouled.

Without hesitation he continued backing and dragged his anchor seaward until in the clear, where he hove in and re-anchored on a safe bearing from Arguello. This position was secure as far as the Farragut was concerned. However, it was so great a distance from the wrecked ships that, in the high surf and variable visibility, it was considered too risky to attempt sending assistance to the disabled destroyers until morning. Had the Farragut's whaleboat not been damaged, it might have been a different story.

There was nothing to do for those aboard the Division Flagship but to stand by, to pray, and to hope for the survival of comrades whose luck appeared to have run out. Quick examination of the Farragut's hull and compartments showed that her grounding and collision had produced some minor lacerations. At the worst, however, she had only a salted-up boiler feedwater tank and a leak in the forward engine room. Both of these casualties could be handled readily without immediate docking. On the whole, her luck, like that of her famous namesake—and based on the same qualities of judgment and determination—had been marvelous.

DD #297—2104:00 hours (Fuller)

On the bridge of the crippled Fuller as she grazed the Farragut's side, Lieutenant Commander Walter D. Seed, her commander, had been doing everything within his power to avoid the impending collision. Too many things had happened too suddenly aboard the Fuller for well-spaced chronological accounting. Short minutes before she struck the Flagship, while backing to kill her headway, a slight tremor ran through the ship.

To Captain Seed, it had seemed quite similar to the distinctive vibration set up when reversing propellers first take hold of the water. There was no warning shock. But, unknown to him at the time, her side had been badly ripped.

"What was that?" the Captain asked as he turned to Lieutenant Homer B. Davis, his Executive and Navigating Officer. Before Davis could make a reply, the attention of all hands on the bridge was attracted by a flare thrown up by a torching stack. It came and went with the speed of light entering through the shutter of a camera lens.

But it lasted long enough to reveal, to those on the Fuller's bridge, a destroyer just ahead with lights that grew from dim to dark.

"All back, emergency," ordered Seed.

He had barely given the order when water was reported pouring into the steaming fireroom, killing the fires. This was a critical blow. In less time than it takes to draw a deep breath, all power went off the ship. Although there could have been little speed on either ship, the Fuller was drawn toward the Farragut as irresistibly as a needle is pulled by a magnet. Threat of collision at sea, with all its potential dangers, is always a terrifying thing—enough to chill the spine of the most stout-hearted seaman.

But that night off a lee shore, lashed by thundering swells, with disaster and death all about them, the thoughts and emotions of those two crews, as the Farragut and the Fuller surged toward each other, undoubtedly will remain with them so long as they live.

Fortunately, the Fuller was pointed in such a way that her bow struck only a glancing blow. With the crunch of metal and snapping of stanchions and stretchers, she was brushed aside and drifted away, a helpless plaything of the pounding seas.

While those on the Fuller's bridge wondered exactly what kind of hell had broken out below, the men in her steaming fireroom fought for the life of the ship and for the lives of those aboard her. A well-honed rock had carved through the port side of #1 fireroom and made a hole that measured about 4 feet in diameter. Through this gaping wound, sea water rushed into the room with the force and volume of water spouted from a gigantic pressure hose. Water tenders and firemen, powerless to do anything to stem it, scattered before this hydraulic blast. Burners were extinguished.

But steam pressures remained high as the icy water rose menacingly around boilers filled with superheated, explosive steam. Even as the screws ceased turning and silence, as well as darkness, descended upon the impotent destroyer, Captain Seed called out to the forecastle hands to get an anchor ready for letting go. Under the circumstances, there was nothing else to do. There was no spot of safety to head for in this multiple-pronged turmoil. He could only hope that his anchor would hold.

On the heels of the disastrous news from the fireroom came Chief Machinist's Mate C. M. Hunt with word that the after-engine room had a great gash where a pinnacle rock had sliced through the Fuller's metal hide.

The engine room, Hunt reported, was taking water rapidly. As this torrent of heartbreaking information engulfed him, Dudley Seed's thoughts flew to the boilers and the danger of their explosion.

The Engineer Officer, Lieutenant (jg) Ralph D. Sweeny, however, was on the job, and the screaming rush of steam from the stacks told the Captain that the safeties had been lifted by hand. But the feeling of relief brought by this reassuring report was to be short-lived. Studying the situation from the starboard bridge wing, Captain Seed was stunned by the grim sight of a string of large and small exposed rocks at close quarters.

Due to the drawing down of the drapery of fog and smoke at the moment, he could barely make them out. He knew instantly that only a miracle could save the Fuller from being flung upon their pinnacles.

The wind and the sea on the Fuller's port quarter were forcing the helpless vessel toward the surf-swept rocks. At the moment when Captain Seed caught a glimpse of the danger that confronted him on the starboard hand, the outlying pinnacles were only some 40 feet away.

Although the anchor had been reported ready for letting go, the Fuller's skipper decided against using it just then, to prevent the destroyer from being thrown on the rocks broadside to the seas.

Less than a minute later, wind and wave brought the Fuller up on the outermost string of exposed rocks. She started pounding immediately.

The ship listed and settled rapidly. She was headed east. Soundings taken on the port side under the stern and the galley deckhouse showed 8 and 5 ½ fathoms respectively. In a matter of a very short time, the ship's bottom was pounded open and she settled with a 45-degree list to starboard. This meant that her starboard rail was awash.

To Seed, who throughout this swiftly paced scene had remained on the bridge, Lieutenant Davis reported all hands were on deck with life preservers on. The men were brought forward of the galley deckhouse. Below, all compartments were filling rapidly.

Now it seemed that time, which had been galloping by, crawled past on leaden feet as the elements made sure of their prey by slowly, ever so slowly, working the destroyer ahead until her bow was jammed between two rocks.

Despite the violent poundings of the ship, Machinist's Mate First Class Frank Moon—and he was really First Class—had been making a survey of the situation on his own account. On the starboard quarter, he had found several fairly large rocks. One of these, about 15 feet off and just aft of the bridge, was studded with grotesque spires.

Frank Moon obtained a strong but light length of line. One end of this he fashioned into a lasso with a large, easily sliding loop. After a few unsuccessful tries, Moon let the loop fall over a large and solid-looking stump of rock. Pulling the loop taut and securing his end of the line around the lower part of the mast, Frank calmly crossed the rope hand over hand. He gained the rock and found that, in a pinch, it could provide refuge to some 20 men.

But Commander Seed, after giving the matter thought, decided that the rock was too small and unsafe. He ordered Moon to return to the Fuller, where he thanked him for his initiative.

While Moon staged his risky but highly courageous bit of exploration, the stranded Woodbury was discovered 100 yards to the east. The Fuller, in fact, had come to grief on one of the outriders of Woodbury Rock.

Messages exchanged between Radioman Novak on the Woodbury and Albert S. Douglas, Radioman 2nd class, aboard the Fuller, on battery operated sets, provided information that the Woodbury, while stranded, was not unsafe.

And that, if she had to be abandoned, her crew had ready access to the rugged wilderness of the desolate little island against which her bow was held in a viselike grip by pinnacle rocks.

The only problem was: Under prevailing conditions of uncertain visibility, crashing breakers and wind-swept seas, how could those aboard the Fuller reach the Woodbury?

She could not launch boats from her up-tilted port quarter. As for putting them over her starboard side, they would be ringed in by dangerous rocks surrounded by wildly swirling waters.

The Woodbury and relative safety were only 100 yards away.

And yet, as far as reaching this haven from the Fuller was concerned, the distance might as well have been 100 miles.

DDs #298 and #301—2105:00 hours (Percival and Somers)

The Percival (298) and the Somers (301) were streaking along at 20 knots when they followed the Fuller around the bend and about 250 yards apart. The Percival, commanded by Lieutenant Commander Calvin H. Cobb, was somewhat to port and 300 yards astern of Dudley Seed's destroyer. In all probability this, plus the benign influence of Lady Luck and good seamanship on the part of "Granny" Cobb, saved her from joining the cavalcade of pile-ups staged by her Squadron mates, one by one, after they made the fatal turn. A few minutes after the Percival had steadied on her new course, a ship ahead, only dimly seen, was observed to sheer sharply to port. At the same time, the Fuller's speed light showed red. "Granny" Cobb at once rang down both engines. As the Percival slid forward, a flash like that of a torching stack was seen quite a distance ahead. Not liking the looks of things, and being safely to port of the upcoming Somers, Captain Cobb sounded his whistle, changed his speed light, backed full speed, and worked the rudder until the ship lost headway.

The first flash of Percival's searchlight revealed the Young on her side and gave a glimpse of the stern cliffs of Bridge Rock on the port beam. Without standing on ceremony, the Percival high-tailed it out of there as rapidly as her full-reversing propellers could take her and on the course on which she came in: zero-nine-eight, true.

At that time, according to her skipper's estimate, she was some 500 to 600 yards from the Young. While the Percival was backing, the Somers and the Chauncey streaked past her on the starboard hand. After getting clear, Captain Cobb anchored in 23 fathoms of water, bearing 282 degrees from Point Arguello. It was only later that he was to learn the full facts about the narrow escape of his ship from the jinns of Honda.

Captain William P. Gaddis, skipper of the Somers, was in his emergency cabin abaft the bridge with Lieutenant Jens Nelson, Exec. and Navigator, when his destroyer rounded the turn. They were figuring the squadron's new course and were disturbed because it did not seem to tally with their own estimates of their position. Nelson's dead reckoning seemed to take the squadron closer to shore and higher along the California coast than the Delphy's navigator did. When, about 2105, the Delphy's frantic "Ships nine turn" signal was received, Capt. Gaddis instructed Ensign John P. Womble, who had the deck, not to make the turn on receiving the order to execute. The skipper's hunch, that something was terribly wrong, was justified a breath or two later when Lieutenant Nelson barged out of the emergency cabin and unceremoniously shouted: "We'll hit the beach in another few minutes if we stay on this course."

The truth that there was trouble abroad was shown on the heels of this warning when Captain Gaddis saw a number of ships ahead; all out of column and some of them backing. Ensign Womble rang up "Stop" on both engines and began to sheer out to port. Gaddis immediately countermanded the OOD's order, called for one-third ahead on both engines, and ordered right rudder. As the Somers swung to starboard and took a westward heading, Gaddis drew an easier breath and thought himself clear of all trouble. But escape was not to be that easy for the Somers. Almost at once, he beheld a large breaker on his port beam. Emerging suddenly from the fog that had been hiding it, this foam-topped harbinger of a submerged reef was only 50 feet away. Although he at once reversed engines to full speed and did all within his power to brake the Somers' forward rush, the bow struck a hidden ledge before the ship could gain sternway.

As the destroyer backed, the port screw struck another hidden obstacle. Although the tips of the three blades were bent and broken, the propeller continued to function. The damage forward consisted of the flooding of five watertight compartments in the double bottoms. Bad!

But not bad enough to compel the Somers to seek shelter in safe anchorage. Throughout his long sea-going life, Captain Gaddis has congratulated himself on the action he took that dark and foggy night. Being an even-numbered ship in Divisional Formation, the Somers, under ordinary conditions, should have sheered out to port when the column ahead spread out like a fan. But his belief that she might be near the beach made Gaddis take radical action in sheering out to starboard. And, as he said at the time: "In doing so, I believe that I saved my ship from sharing the fate of the other destroyers in Division 33."

DD #296—2107:00 (Chauncey)

When the Chauncey, tenth destroyer in column and the last vessel in Division 33, surged from the haze of the open sea into the opaque realm of the coastal fog, Lieutenant Commander Richard H. Booth, her Captain, had neither suspicion nor knowledge of the maritime Gotterdammerung that awaited him. Thick gray fog, marbled by streaks of heavy black smoke, swirled in slowly moving masses over inky waters edged with ghostly white spume. Only a few swift glimpses were allowed him of the somewhat obscured scarlet breakdown or backing lights that showed on left and right.

Straight ahead was a vessel whose flashing white speed light revealed it to be moving forward at two-thirds speed. Captain Booth immediately relieved Ensign Robert Greenwald of his responsibilities as Officer of the Deck and went to two-thirds speed. He also warned Machinist's Mate C. G. Ostergard, who had the engine room watch, to stand by for fast action on the engine telegraphs. From the radio shack, Radioman Frederick Fish reported receipt of a message from the Delphy warning to "keep clear to westward."

But no execute ever followed. Dick Booth's immediate reaction was that there had been a collision ahead or man-overboard. He rang down his engines. As the Chauncey responded slowly to a quick succession of heavy right and left rudders to reduce its forward speed, the appalling sight of the Delphy smashed against

Bridge Rock on his starboard bow with the Young capsized on the starboard quarter met Booth's astounded gaze.

Ahead, revealed by the Chauncey's searchlight, lay rocks and cliffs. It was an unbelievable situation, but in Dick Booth's clear-thinking mind there were two things he had to do: get his ship to hell out of there and try to rescue the crew of the Young. By the weird light of a flaming carbide bomb that floated on the water, he saw that scores of men clung with the strength of sheer desperation to the port side of the overturned ship. Something must be done for them.

On the Chauncey's deck, Ensign F. A. Packer's reaction to the emergency was the same as his skipper's. Swiftly, he seized a whaleboat fall and hove it across the narrow gap to the Young, but unfortunately lost the end of it. From the capsized hull of the Young, Bill Calhoun had been watching the oncoming Chauncey with increasing anxiety. He feared that, if the other destroyer should even sway the Young with her propeller wash, his disabled ship might slide off the ledge on which it rested, sink into deep water, and jeopardize the lives of all hands.

"For God's sake," he shouted as loudly as he could, "don't ram us! You'll knock us off this rock!"

Captain Booth did not hear the appeal. Now the Chauncey was close enough to the Young so that those aboard the former could hear a chorus of high-pitched voices shout: "Don't hit us! Don't hit us!"

Meanwhile the Chauncey, backing full speed, was desperately trying to claw off the lee shore and away from the cliffs, rocks, and wreckage whose touch would be the kiss of death. On his way out, Booth planned to make another try for the Young. But, alas for his heroic plan, Fate willed it otherwise.

Down in the Chauncey's engine room, Electrician's Mate Ostergard and his gang stood alertly at their stations. Word had come down through the hatch of the plight of the Young and the impending rescue of her men.

If the men in the engine room could do anything to speed and ensure their rescue, they could be depended on to do their best, come what may.

The determination of these resolute souls was to be subjected to an immediate test. The Chauncey's stern, seized by the outgoing undertow, was raked against the port propeller blades of the Young.

The sharp tips of the screws cut through the Chauncey's starboard side as if it were made of cardboard. Ostergard and his men stared with fascinated eyes as metal blades from nowhere cut through the engine room's steel plates as easily as paper-knifes opening envelopes. Ostergard lost no time staring at the spectacle, but reported the damage to the bridge where Captain Booth was unaware of the disaster. But further action was taken out of the skipper's hands when his destroyer's bow was driven to the right by a series of knockout, heavyweight breaker blows. By now, water stood waist high in the engine room and the Chauncey suffered immediate loss of all power. The battering blows of the breakers reached their peak when a giant wave tossed the destroyer on a sunken reef. As engine and boiler rooms were secured, Captain Booth ordered Chief Boatswain's Mate M. Frye to let go both anchors.

Because the Chauncey was almost dead in the water at the time of the accident, she ran on the rocks with less force and injury than had been the fate of her sister ships. The vessel grumbled and groaned as waves snapped her back and forth, but Captain Booth hoped that, come high tide, she might be able to float again.

This hope was drastically reduced when Booth took further inventory of his position. Minute by minute, the waves pushed the Chauncey closer to the cliffs of Bridge Rock that towered on her port side. Although the destroyer seemed sound enough to float, there was the very possible danger that, with the change of the tide, she might be thrown against the cliff and jeopardize the lives of all aboard.

"Pass the word," said Captain Booth to Lieutenant (jg) Clarence V. Lee, his Exec, "to stand by to abandon ship."

As "Chink" Lee turned to execute the order, Captain Booth measured the shortening distan
ce between the destroyer and the precipitous side of Bridge Rock. Between it and the ship ran a swirling maelstrom of waves that poured like flash floods through a narrow canyon. It would, Booth knew, be a man-sized job to get his men safely across this seething vortex. But it had to be accomplished.

Destroyer Division 32

DDs #306, #307, #302, and #305—2105:00 hours
(Kennedy, Paul Hamilton, Stoddert, and Thompson)

The Kennedy (306)—flagship of Commander Walter G. Roper, ComDesDiv 32, and commanded by Lieutenant Commander Robert E. Bell—was barreling down the Chauncey's frothy wake with the three destroyers (of the five in the Division) which took part in the 20-knot run.

In the Kennedy's wake came the Paul Hamilton (307), commanded by Lieutenant Commander Tracy L. McCauley; the Stoddert (302), commanded by Lieutenant Commander Leslie E. Bratton; and the Thompson (305), commanded by Lieutenant Commander Thomas A. Symington.

The Reno, one of Division 32's destroyers, had, as will be remembered, been assigned by Rear Admiral Kittelle to make a full-speed engineering run from San Francisco to San Diego. This run, as previously related, was disrupted when Captain James R. Barry, her skipper, abandoned the speed test to pick up survivors from the Pacific Mail steamer Cuba stranded off San Miguel Island. Incidentally, the grounding of the Cuba was to have a considerable bearing on the destiny of Destroyer Squadron 11 on that particular day. When the long-delayed news of the Cuba disaster became known, Squadron 11 was already well past the Golden Gate's Potato Patch and the Farallones.

Flashes of the disaster were picked up in the Kennedy's radio shack during the afternoon and passed on to Commander Roper, a blunt, salty, and colorful character of the old school.

By ship-to-ship radio phone, Roper requested Captain Watson for permission to take Division 32 full speed down to the scene of the accident.

The suggestion was not accepted by Captain Watson. He believed that Division 32 would reach San Miguel much too late to be effective.

In this assumption, although he did not know it at the time, Captain Watson was correct. Captain Barry and the Reno were to perform a full-out rescue service.

But Commander Roper, with the bulldog tenacity for which he was famous, kept on insisting. Watson, who had a streak of stubbornness of his own, kept on declining and finally told Roper off in no uncertain terms.

The Squadron, he asserted, was on an important engineering run to determine if 20 knots could be maintained over a stretch of many hours. The run would be executed. Period I, Roper, easily ruffled, hung up in a huff. He never wore his blue uniform coat buttoned so tightly but that his rambunctious disposition had a free flow beneath it. Following his telephone conversation with the Squadron Commodore, the irate Roper retired into his cabin to huff and puff and blow the man down.

Professionally, Walter Roper was not generally considered a success. Still only a Commander, he was actually senior to Captain Robert Morris of the class of 1900, whose Division 33 led Squadron 11's formation. As "anchorman" of the class of 1898, Roper was an exception to the rule that a Midshipman on the bottom rung of his class is very apt to make flag rank. As a man, Commander Roper was filled with frustrations and not too easy to get along with. He made critics where it hurt. As a sailor, he was well above average; a strict disciplinarian who ran a taut ship or division; a commander who knew what he wanted and usually got it. As one young officer recalled, "He sure straightened out DesDiv 32."

At 2000 hours, Ensign Sampson G. Dalkowitz relieved Ensign Max Welborn as OOD. As is customary, they went over the chart together, and at this time Welborn ventured the opinion that they were making good a course of 148. In other words, the Kennedy was to the east of 150, true. Further, Welborn believed that, because of difficult steering due to following seas, the squadron was not making 20 knots over the ground. To Dalkowitz's poorly disguised alarm, estimates based upon these beliefs showed that the destroyers could be heading for the beach. The two ensigns (young, full of beans, and eager beavers) recalled that on other occasions, in passing up and down the coast, the Kennedy had encountered inshore sets of current. During this discussion, "Baldy" Bell, the skipper, had come on the bridge. When the ensigns tried to call their conclusions to his attention, "Baldy" waved them off with the remark that the Delphy knew what she was doing. When, around 1800 hours, Sparks picked up a radio bearing of 320 that checked with the chart projections of Lieutenant R. F. MacNally, Exec, and Navigator—the ensigns reported their fears for the safety of the formation to Division Commander Roper.

He listened with thinly disguised lack of interest. Then, with the cutting sarcasm he loved to employ, Roper observed: "Seems like you two think you know more than the Squadron Commander!"

Then, perhaps regretting his acid remark, Roper added with less sting in his voice: "The Commodore gave me hell about the Cuba, so I don't want to stick my neck out unless we are sure that we are right."

Roper made off as if he were about to leave the bridge. Then he halted, turned slowly, and, smiling wryly, said: "All right, if you're afraid, get over to the right."

Dalkowitz, who now had the deck, gave orders that placed the Kennedy some 200 yards to starboard from the line of formation and opened the distance between the Flagship and the Chauncey, which was the nearest ship ahead, to 350 yards.

Astern, the other ships of Division 32— the Hamilton, the Stoddert, and the Thompson—swung to the right, and away from the hidden dangers of the nearby shore, as they reached the Kennedy's turning point. It is always idle to speculate on what might have been, but to this day Dalkowitz believes that, had it not been for the bawling out the Division Commander had received earlier that day from the Squadron Commander, Roper would have voiced his doubts to Captain Watson and all the ships in the squadron might have been saved. It was the old, sad story: For the want of a nail ... a kingdom was lost. About 2100, the Kennedy, by radio phone, received a signal to take a zero-nine-five course on reaching the squadron's turning point. Calling the Captain and "Mac" MacNally into quick conference, Roper delayed execution of the order.

Lieutenant MacNally explained that, at 2058, a radio compass signal, giving the bearing from Point Arguello as 323, had been intercepted. As against the 320-bearing given in the 1813 signal, this would indicate that the Delphy's position was to the north of Point Arguello and closer inshore.

In other words, she was not clear of the California coast and free to enter Santa Barbara Channel. Watching from the bridge, the three senior officers observed that, in a few minutes, there was a noticeable bunching up among the ships ahead. At 2105, the Kennedy slowed. Soon after, she came to a stop, as did the other ships of Division 32. At that time, the Chauncey was on the Kennedy's port beam and about 1000 yards ahead.

As the destroyer, dead in the water, lay tossing on the restless waves, a bump was felt in the bow. It could have been caused by hitting a log, a whale, or a reef. "Holy Moses! We're aground," shouted Captain Bell. "Back. Full."

After a minute or so of backing, which pulled the destroyer to seaward, "Baldy" rang down his engines and took soundings.

"By the mark seven!" sang the man at the lead. But even before he called out, "Baldy's" sharp eyes had caught sight of the tell-tale red rag affixed to the lead-line at 7 fathoms. It gleamed wetly, like fresh blood, in the beam of the bos'n's electric torch.

The sight brought Bell a shock of surprise. The Kennedy and her sister ships had, indeed, been closer to the shore than they had believed. Five distinct curves or gradients, marked in pale blue, run along California's coastal line on the Hydrographic Chart.

First comes the 50-fathom curve, then the 40-, the 30-, the 20-, and, lastly, the 10-fathom curve. This latter curve is nearest to shore and therefore the closest to danger. For easy recognition, it is marked on the chart in a deeper blue than the curves beyond 10 fathoms.

In some spots along the coast, the spaces between the individual curves are fairly wide and, as a group, may extend several miles to sea from a gradually shelving coast. But off Point Pedernales the belts are narrow and closely bunched. Thus a ship off its course may quickly make the transit from safe depths into the dangerous waters off Honda. This unpleasant situation had barely had time to register its impact when Sparks reported receipt of the signal "Nine turn" from the Delphy.

"Disregard it," commanded Roper to the Kennedy's skipper.

Using whistle signals and megaphone, he ordered the ships of his division to disregard the order to make the suicidal 90-degree turn to the left. Instead, they were directed to turn to the right and seaward. The Stoddert, not getting the message, charged up close aboard to, and seemingly bent upon passing, the Kennedy. But she was quickly brought to heel by Roper's

loudly megaphoned: "Get to hell out of there: fall in astern!"
Again the Kennedy showed backing on her speed light and went
astern Emergency full. The action was precipitated by the sight of
unaccountable but widespread confusion ahead and the sound of
whistles and sirens. On her bridge there were comments about
the likelihood of man-overboard or collision. But back in the
minds of all who saw the prelude to chaos was the fearful dread
that ships were stranded on the beach. Making sternway into the
seas and at Emergency full, the Kennedy moved so fast that her
fan-tail was practically submerged. Twice those aboard felt the ferocious, battering-ram impacts made by heavy seas crashing
against the vessel's stern.

These onslaughts caused many of her complement to believe
that the destroyer had tangled with a rock or a reef. Feeling her
way, by means of frequent slow-downs and the use of the sounding lead, the Kennedy eventually found herself and her charges in
safe waters at 90 fathoms.

Guided by soundings and radio bearings, Division 32 spent
the night standing up and down the coast off Point Arguello on
north and south courses. By piecing together the weak and frag-
mentary radio signals sent out by stranded ships that still had
battery power, Roper and his skippers were able to form a faint
picture of the tragic fate that had overtaken their squadron mates
and which they, by the narrowest of margins, had been spared.

Destroyers into Derelicts- Seven DDs on the Rocks—2105 to
2110 hours

The hour was now drawing toward 2110:00. Only about 5
minutes had elapsed since the Delphy tried to send out her futile
warnings. One by one, nine ships following in swift succession,
according to long-standing concepts of destroyer doctrine, had
steered duck-trailing courses in follow-the-leader fashion toward
the hell that awaited them behind the curtain of fog that hid the
rocks of Honda.

In an area, hardly more than 500 yards long from east to
west and just about 800 yards wide from north to south, seven
proud destroyers had gone to their deaths and two more had felt,
but escaped from, the fangs of the Devil's Jaw. For mere minutes,
in some instances only seconds, whistles blew, sirens screamed,
searchlights swung and blinkers gave their warning. Then, as
power ran out, all grew dark and silent, the silence of a graveyard

in which seven stranded destroyers stood as immobile as potential coffins for almost 800 courageous fighting men—men who hoped against hope that a miracle of rescue might be wrought before complete disaster overtook them all.

At that time, from Squadron Commander to the lowest rated mess attendant, not a soul knew where he was. But most of them feared that they had hit outlying rocks of desolate San Miguel Island. The very island where, as they had learned on scuttlebutt from their radio shacks, the Pacific Mail Steamer Cuba had been reported stranded in the hours just before dawn that very same day.

Meanwhile they were not downhearted. The Lord, they knew, would find a way. Many, with prayerful hearts, breathed through numbed lips the words of the hymn which has been sung in Navy chapels since earliest times:

Eternal Father, strong to save,
Whose arm hath bound the restless wave,
Who bidd'st the mighty ocean deep
Its own appointed limits keep:
O hear us when we cry to Thee
For those in peril on the sea.

6 - GIORVAS SMELLS SMOKE

IN THE CITY OF SANTA BARBARA (60 miles to the south) and in
the town of Lompoc (which lies an L-shaped stretch of 15 miles to
the north of Honda) it was Saturday night and a time for making
merry after the week's hard work.

But at the Railroad Section House on Honda mesa, it was
just the end of another day and the beginning of another lonesome night.

Being, by habit, a man of solitude who fended for himself at
housekeeping tasks with ease and simplicity, John Giorvas had
cooked and consumed his dinner, cleaned up his kitchen, and lit
his post-prandial pipe. For the better part of the evening, he had
been toying with a newly acquired crystal radio receiving set. But
either the radio or the reception was bad and, eventually, John
gave up listening to the raucous voices of the spheres.

The hour was just past 9 o'clock when Giorvas decided to call
it quits. He was tired. As always, it had been a long, hard day. He
blew out his living-room lamp but, before he headed up the narrow flight of stairs to his attic bedroom, Giorvas decided to step
outside for his nightly look at the rails, the sky, the mesa, and
the sea. The streamers of fog that had been drifting together just
around sunset had now closed in. Overhead, neither moon nor
stars could be seen. The side-visibility was also bad. Looking up
and down the rails that ran before his door, John could count
barely half a dozen ties on either hand.

The night was raw and the northwest wind had a cutting
edge. The sea was, of course, invisible. But the well-timed beat of
the surf and the occasional crash of breakers reached his ears.
He listened briefly and with evident enjoyment. It was music.
Driven by the chilly dampness of the fog, Giorvas re-entered his
house and fumbled his way up the stairs to his tiny bedroom.
John was just about to strike a match to light his night-lamp,
when he froze in mid-passage. His eyes, well adjusted to darkness, had caught a flash of light to seaward through the window
pane. Also, he had heard something from the direction of the sea
that sounded like a crash.

What could that be?

Instead of striking a light, Giorvas opened his bedroom window. He looked across the rails and the mesa; stared and listened; his senses of sight and hearing as taut as violin strings. All he heard as the minutes sped by was the familiar pounding of the surf. Then his eyes caught another glimmer of light. It shone briefly and mistily through rents in the slowly flittering curtains of fog.

Now, also weak and stifled, as if they came from far away, his ears picked up a ghostly mixture of shrill and deep sounds. They had an odd, unearthly quality. Filtered as they were, through countless folds of fog, they seemed vaguely like voices lifted in mortal distress. What could that be?

To John Giorvas there was but one answer to that question: A ship was in trouble. Those shrill voices, like the screams of banshees, he could not place. But the deeper notes sounded like the frantic blasts of a steamship whistle.

A ship in trouble—and a big ship at that. The flashes of light he had seen seemed to have come from a spot just beyond Point Pedernales. Grabbing a lantern from the storeroom off his kitchen, John barged out of his house at high speed, ran across the rails, and headed down the sloping mesa toward the coast. The straight-line distance between the Southern Pacific track and the edge of the mesa is about half a mile, give or take a few dozen yards.

In the dark, even assisted by the light of his lantern, the going was difficult, not to say dangerous. The entire expanse of the mesa is closely pitted with holes dug by rodents and other ground-dwelling animals. At night, a man, even moving at a half-run, could easily stumble into one of those holes and break a leg. Other hurdles in John's path were the ever-present spreads of tangled matting formed by roots and branches of desert vegetation. Try as he might, Giorvas could not make the distance on the run. He had to pick his way, and with care.

This procedure took time and made John's tongue burn with a flame of highly vocal impatience. To make time, Giorvas decided not to stumble across the mesa in a straight line, but edged toward his right, almost paralleling the tracks. His purpose was to reach an old wagon road that ran from Canada Honda—at a point near the railroad trestle that crossed the canyon—to a gravel quarry near Saddle Rock atop Point Pedernales. It was a less direct route. But once on the wagon road, John made better time. Still, even under those circumstances, it took him a good 10 minutes to reach the top of the cliff that opened upon the Devil's Jaw. As he stood upon its edge, all John could hear was the sea; all he saw was the fog.

Throwing himself flat upon the ground, he reached head and shoulders out over the edge of the cliff to discover what—if anything—was happening on the invisible waters in the Stygian darkness some three-score feet below. Straining his ears—in the brief pauses between the crashings of the surf—he thought he could hear voices. Willing himself to see through the murk, he believed he could catch occasional flashes of light. They seemed to flicker like restless beams from electric torches. Suddenly his nose, sampling the moist air that came swirling up along the cliff, made a vital discovery. The cold, damp current was heavy with the odor of smoke. John knew it well, the acrid smell of the greasy black smoke of oil-burners.

Now John knew the answer for sure. A ship was stranded off Honda. And, he guessed, not far from the spot where the Santa Rosa had met her death only 12 years earlier. Again, he thought he heard voices. "Ho, there!" he yelled at the top of his lungs.

No reply.

Again he shouted. This time he filled his lungs to the bursting point and leaned as far over the crumbly edge of the precipice as he dared without losing his grip or his balance.

"Ho, there!"

No reply.

But, as if it were some sort of an answer, a flaming light seemed to swim into Giorvas' vision. It was, in all probability, a calcium flare attached to a free-floating life preserver. The fog may have lifted momentarily or the tall, hot flame of the flare could have burned a hole in the moist masses of vapor. At any rate, for one brief moment, John beheld the four stacks of the stranded S. P. Lee—although he did not then know her name— silhouetted against the seaborne light. He saw the bridge structure and the hefty 4-inch gun on her foredeck. Having seen this, John had seen enough.

Help was needed, but more help than he was able to supply with his limited resources. His first job was to flash an alarm to Surf, 5 miles up the line. His next, to return to the cliffs of Honda with his section crew and lend a hand as best he could. Once more John stumbled through the darkness, often plunging forward as a foot was caught, but always recovering his balance before he actually fell. He ran as fast as he could. But this time, what with facing a fairly steep slope, the going was harder. Recalling that his house was dark and that he had seen no light in the section bunkhouses, Giorvas did not cut across the mesa from the poorly defined wagon road. Instead, so that he would not waste time trying to locate the darkened Section House in the all-enveloping fog, John made a beeline for the railroad track and ran between the rails to his door.

Without taking time out to light his lamp, John cranked the handle on the railway telephone that connected him with the ever-open dispatcher's office in the Southern Pacific's station at Surf. When telegrapher William Pittman, the second-trick operator at Surf, answered the call, Giorvas shouted: "This is Johnny Giorvas. We have a wreck up here at Honda. I think that a big warship has gone on the rocks. Is Mr. Maes around?"

W. J. Maes, General Foreman of this part of the railroad's coastline, was John's immediate superior.

"No," replied Pittman. "He was here a while back. Said he would be at his house if anybody wanted him."

Then, as the nature of the call dawned upon him, Pittman asked: "But what's that about a warship on the rocks, Johnny?"

There was no reply. The wire had gone dead as Giorvas hung up. This was no time for small talk. Despite Pittman's frantic rings, there was no answer from Honda. The energetic little Section Foreman had turned to the nearby wall instrument of the Santa Barbara Telephone Company. As he lifted the receiver, he discovered that—as usual—the heavily populated rural party line was busy. And the two women who were talking obstinately declined to yield their rights of free speech to any man for any reason. Minutes slipped by before the wire was open and John had a chance to inform his superior about his discovery.

"A warship, you say," commented the unruffled Mr. Maes whose very job was to thrive on emergencies. "Okay, Johnny! You rouse your gang out and go down to Point Pedernales with lanterns and ropes. As soon as you know what's what, call me back. I'll be in the dispatcher's office so you can get me on the Company line."

"Okay, boss, okay," agreed Giorvas as he hung up. Next he loped out of his kitchen door and struck resounding blows on a steel triangle with a steel bar. The resulting sound waves sped through the night and into the bunkhouses where they awoke section hands from their slumbers.

The signal was an emergency call, so the men who had remained on hand during the weekend, about a baker's dozen, dressed hurriedly and reported to Giorvas. Handing them lanterns and bidding them carry such lengths of rope as they could lay hands on, Giorvas led his gang across the mesa to Point Pedernales to give whatever aid they might be able and to obtain further details about the ship that had come to grief in the Devil's Jaw of Honda.

Maes Spreads the Alarm

In the seaside village of Surf, General Foreman Maes hurried to the Southern Pacific Station and entered the dispatcher's office where Telegrapher Pittman's key was clicking steadily.

"I'm trying to raise Trainmaster Foley up in San Luis Obispo," explained Pittman.

"Fine," endorsed Maes. "Hope he can come to the key so that I can talk with him."

After a few minutes' wait, Trainmaster Foley was reported as present in the dispatcher's office at San Luis Obispo. The latter city, an important Southern Pacific division point, lies 51 miles north of Surf.

Following a brief conference in Morse code, during which the two operators acted as intermediaries, Mr. Foley took quick action. He rounded up Foreman Tobin of the steam shovel gang and Foreman Motis of the fence gang. They were instructed to proceed at once, with their men and equipment, from San Luis Obispo to Honda and give whatever assistance they could in the rescue operations. Knowing the layout at Honda, Trainmaster Foley rightfully felt that teams of men who were expertly trained in the handling of rope and tackle could be of vital assistance.

At the same time, he sent a message to the Southern Pacific's main offices in San Francisco. There, Night Chief Dispatcher A. Keller relayed the news to the Marine Exchange and to the Duty

Officer at the Mare Island Navy Yard. But, since it was believed that the report had been garbled in transit and that it really referred to the wrecking of the Cuba off San Miguel, the immediate reaction was to brush the news off as unimportant.

At this hour, not even the San Francisco newspapers or wire services were aware that the greatest and most dramatic naval disaster in modern history had been staged 217 miles south of the Golden Gate.

As he waited at Surfs railroad station for further news from Giorvas, it occurred to Maes that any stranding dealing with a warship—maybe even a battleship—demanded better attention and investigation than mere railroad men could offer.

Reaching a quick decision, he set out to contact City Marshal W. S. Bland, head of police in Lompoc, some 9 miles to the east of Surf. Getting Bland on the telephone that Saturday night proved a difficult undertaking. The town was literally thronged to the rafters with visitors who had come in from all over the surrounding district to attend Lompoc's annual harvest pageant known as The Valley of Gold Fiesta. The gold in Lompoc Valley is the greatest outpouring in the world of flower seeds from hundreds upon hundreds of acres of fields of flowers. These seeds are sent all over the globe for commercial distribution. And in Lompoc, a successful seed harvest was, and is, worth a bang-up celebration. At long last, Maes located City Marshal Bland at a dance that was staged as the opening gun of the pageant. He was standing on the sidewalk outside the building in conversation with Motor Traffic Officer Glenn Baker and Ronald L. Adam, owner and editor of The Lompoc Record.

Adam, always an enterprising reporter with a nose for news, was getting data for a story about traffic jams, when Bland ran off to answer a telephone call.

"That was Walt Maes, General Southern Pacific Foreman over at Surf," he explained to Adam and Baker on his return. "He's heard that a warship has gone aground up at Honda and wants me to go up and investigate."

"How're we going to get up there?" asked Ronald Adam dubiously.

"Fly?"

"Where do you get that 'we' stuff?" chided Bland. "But," he continued, "if you want to know and if you want to go, Walt'll have a motor track car ready for us in front of Surfs station platform."

"Well, what are we waiting for?" shot back Adam, impatiently. "Want me to drive us to Surf?"

He had a newspaperman's hunch that a big yarn was in the offing.

Editor Adam's hunch was better than he knew. He was standing on the very threshold of a reporter's paradise—for many, many hours he was to have an exclusive world-wide scoop on a disaster story that was to put the name Honda in banner headlines on the front pages of the nation's press not only for days on end but over a spread of many weeks to come. It is a name that will never be forgotten in the United States Navy.

"Come along to my office," offered Bland.

"My car is the fastest bus in town. We'll pick up a coil of tow rope in the AAA garage on the way."

The City Marshal turned to Traffic Officer Baker, saying: "Glenn, you get on your motor bike and clear the way for us. I want to get to Surf in nothing flat." He looked at his watch. Just 10 o'clock. In a matter of a few crowded minutes, Bland had borrowed a tow rope and thrown it into the open tonneau of his ancient but powerful touring car. He started its engine, pulled away from the curb with an explosive roar from the wide-open exhaust, and rolled toward Surf as the speedometer climbed past 60—65 to 70 miles an hour. The county road, full of curves and badly paved, was also narrow. But Traffic Officer Baker did a life-saving job by riding ahead with croaking klaxon and waving cars off the road into the dry and shallow ditches.

"I'm quite sure that I have material for your obituary in my files," yelled Adam to Bland, who was crouched over the steering wheel with an iron grip on its rim. "But, I'm hanged if I have my own."

As Bland replied by pressing the accelerator deeper onto the floorboard, Adam added: "Say, Chief—are we on our way to a wreck or trying to make one?"

The sparse lights of Surf, gleaming feebly through the coastal fog, came into sight around a bend in the road. Bland slowed and, with brakes squealing, came to a skidding stop before the Southern Pacific Station.

He drew his large, old-fashioned turnip of a watch from a pocket. It had stopped at 10 o'clock. "Look," he grinned, "I said we'd make it in nothing flat!"

Abandoning the S. P. Lee

The seas were breaking all the way across the S. P. Lee and one had to be constantly hanging on to something to keep himself from being washed overboard. With a 30-degree list to port it was impossible to launch a boat on that side, where the high-rearing surf roared over her rail. It was equally out of the question to put a boat over the vessel's starboard side because of the vortex of madly churning water between ship and shore.

As it was, the position of the S. P. Lee was perilous in the extreme. But now, from the port wing of his bridge, Captain Toaz saw another potential danger bearing down on his helpless destroyer. Barely visible through the fog and the sheets of spray that whipped across the S. P. Lee's bridge was the outline of a slowly approaching destroyer.

She was obviously crippled and out of control. The vessel was bearing down upon the stranded destroyer s port bow, stern first, and posed a deadly menace. Toaz estimated that she was only about 100 yards off and coming closer. If this vessel crashed into the S. P. Lee, heavy loss of life on both ships could result. With a prayer on his lips, and helpless to do anything to ward off the impending collision, Captain Toaz watched the disabled destroyer drift closer. She was the Nicholas.

But, at that time, "Toze" Toaz had no way of knowing her identity. Nor did it matter. A destroyer by any other name would have been just as deadly. Suddenly, Toaz hardly dared believe his eyes. The drifting ship was no longer driving before the seas.

As if in answer to his prayer, she had come to a halt, caught on a cluster of pinnacle rocks. The skipper of the S. P. Lee drew a deep breath of relief. His ship had been given a reprieve. Doom was still suspended over his head.

But, at least, he had a chance to escape absolute disaster in the event the other destroyer should be thrown free of the rocks that held her. By this time it was obvious that the S. P. Lee must be abandoned before she and all in her were pounded to death.

Leaving the bridge, he worked his way to the galley, just abaft the bridge structure, where a large number of men had sought shelter.

After a hasty conference with his Exec., Lieutenant W. E. Tarbutton, and the Chief Boatswain's Mate, Captain Toaz called for volunteers to take a raft across the wildly churning water to the precipitous cliff which, from the canted deck of the S. P. Lee, seemed to tower to a height of at least 100 feet. From among those who volunteered, Coxswain C. M. Carlson and Seaman First Class C. G. Stahl were selected to stage the death-defying effort. Using a small rubber raft—commonly called a doughnut (for that was what it looked like)—the two men took a stout coil of signal halyard and fought their way across the stretch of water that separated the destroyer and the cliff.

It was a distance of only 50 feet, but every foot of the journey was a seagoing nightmare. The doughnut spun and tilted like a top. They had to paddle desperately to keep it under control. The tiny craft might never have made the base of the bluff had it not been for the surf's willingness to carry it there. On making shore, the raft was bounced against the cliff time and again before Coxswain Carlson could loop a line over a rock and make fast long enough for him and Stahl to jump ashore with the remaining length of signal halyard.

They had to cling, spread-eagled, to the wet, slippery, pitted surface like human flies. To use an old time sailor's phrase—they needed fish hooks in their toes and fingers to hang on. A boat would have been smashed to splinters at the first impact with the wall. Only a craft with the resilience of a doughnut—bouncy as a tennis ball—could have accomplished this exploratory landing. When Carlson and Stahl finally found footing on the face of the cliff, they had to hunt for a boulder or a rocky bollard that was large enough and strong enough to hold the heavy line which they hauled over from the S. P. Lee.

This done, they took the raft on another perilous passage back to the destroyer. The operation, begun about 2130 hours, had consumed about half an hour. Meanwhile Captain Toaz had returned to the bridge of his ship. There, he was soon joined by Captain Robert Morris, Commander of Destroyer Division 33.

The latter had been in the radio shack hoping to obtain reassuring news of the remaining destroyers of his division—the

Young, the Woodbury, and the Nicholas. But Radioman J. H. Travers was unable to maintain contact with the Woodbury—the only destroyer whose spark-set was working—long enough to receive much information before his own radio became useless. Two things he did learn.

The unknown destroyer on the S. P. Lee's port hand was the Nicholas and the ship on the rocks astern of Captain Morris' Flagship was the Delphy, Flagship of ComDesRon 11, Watson.

With a signalman at his side, equipped with a powerful electric torch, Captain Morris awaited his chance, in the port wing of the bridge, to establish signal communication with the Nicholas. After several futile tries, during intervals when the fog curtain lifted, contact was eventually established.

From Captain Roesch, the Division Commander learned that the Nicholas was hard and fast on the rocks but in no immediate danger of breaking up. It was, therefore, decided that no effort should be made to abandon the Nicholas until daylight unless there was a deterioration in her position.

Captain Morris also made plans to establish a signal watch on the shore for the rest of the night. The purpose of this watch was to keep signal communication with the stranded destroyer and to summon assistance if it should become necessary to attempt to get her men ashore before daylight. As Captain Morris and Commander Toaz discussed these arrangements and prepared for their execution under the supervision of Lieutenant (j.g) S. L. Huff, Radio and Communications Officer, Coxswain Carlson (who, with Stahl, had just boarded the ship from their doughnut) reported to Lieutenant William E. Tarbutton, Exec, and Navigator.

To make his voice heard above the crash of seas and the grinding of steel on the rocks, the Coxswain had to shout.

"We have secured the line to a pinnacle on the cliff, sir," Carlson announced, "and I feel certain that it will hold."

"Very well," Tarbutton replied, "then we'll get the ferry started."

"But, sir," cut in Carlson, "it's not as easy as that. The question is, what'll we do with the men once we get them over there? That cliff rises straight up. There's a ledge, it won't hold them all."

"That's tough," said Tarbutton. "There isn't a beach of any sort where they can land?"

"No, sir. Just a narrow ledge of rock. And even at this low tide it's awash. Besides, as far as Stahl and I could see, it doesn't run along the entire face of the rock. There's only one way out and that is to climb the cliff. It looks mighty high and steep, but with your permission Stahl and I would like to try it. If we get to the top, maybe we can find a better place to rig the ferry."

Permission granted, these two determined sailors returned to the rugged face of Honda. As bad luck had it, the S. P. Lee was being driven in against the northernmost, steepest, and highest side of Point Pedernales.

When Carlson and Stahl again landed on the narrow ledge, the Coxswain flickered his electric torch as signal that the doughnut could be hauled back to the ship. The two explorers had, indeed, burned their bridges behind them. To expedite the operation of disembarkation, the doughnut was sent ashore with another line, this time under the command of Ensign William D. Wright.

With two lines available, the destroyer's large Carley rafts were launched and the Captain's order to abandon ship was put into execution. Under Tarbutton's calm and expert handling, the departure was conducted without any haste, confusion, or disorder. The procedure was slow, because Tarbutton played it safe—only eight men per raft per trip. The Carleys reached their destination through the simple process of the passengers hauling them through the hurtling surf by clinging to and tugging on the "ferry" lines that stretched overhead within easy reach.

Last to leave the ship were Captain Morris and Commander Toaz. Just as the skipper tumbled into the raft—where Tarbutton, Steinhagen, and Small were already waiting—the long-drawn call of a deep-toned whistle came through the night. On the ledge below the cliff and aboard the raft, which bobbed like a cork along the side of the now abandoned S. P. Lee, men lifted their heads. Once—twice—three times—the round, full, penetrating clarion call reached their ears. "Good Lord! It's a train!" shouted Tarbutton, his oil-smudged face agleam with relief in the light of a hand torch.

"Yay, yay; hurray, hurray!" came yells from the dozens of shivering men clustered on the rocks. "Let's go to town!"

"Thank God for that," said Captain Morris, fervently. "Thank God we are on the mainland—not on San Miguel Island. Has anyone got the time?"

"Yes, Commodore," replied Commander Toaz, turning on his flashlight, "I have just twenty-five minutes past ten." Then, with a quick laugh, intended to relieve the older man's tension, "and I'm afraid we've just missed the good old Lark, northbound to San Francisco—"

The long-drawn hoots of the locomotive's whistle pierced the wall of fog and were heard—as a welcome and reassuring sound—aboard nearly all of the stranded destroyers. While the S. P. Lee was being abandoned, Coxswain Carlson and Seaman Stahl had probed along the cliff for foot- and hand-holes that would lead them to the top. It was hard work, but at long last they reached the edge of the mesa.

There, by following the sloping top of the cliff, they discovered that, toward the south, the rim fell off to a height of only some 30 feet. Working their way down gingerly, they reached the base and discovered their shipmates literally standing shoulder to shoulder along the length of the narrow ledge. Cold, stiff, tired—and not a little scared. Some of the older and sturdier men had all they could do to keep their younger shipmates, many mere boys, from toppling into the sea. When Carlson and Stahl finally arrived and told of finding a fairly negotiable passage, the men began moving upward. It was a slow, dangerous, and painful climb and no place for the faint-hearted.

The wall of the cliff was not only steep but also lined with outcroppings of lava which through centuries of wave action had become as sharp and pointed as the shards of broken bottles. Since the cliff offered no other holds for the climbers, they often cut deep and bleeding wounds into bare hands and naked feet. Few reached the top without blood flowing from cuts sliced into the flesh of half-naked men who, too fatigued to continue their climb, leaned their bare bodies against the razor-edged rocks for moments of respite from their tormenting efforts.

Destroyers of that day did not carry regularly assigned medical officers among their personnel. However, the Navy Medical Corps was well represented aboard by highly skilled and exceptionally resourceful pharmacist's mates. Many of these were really very able doctors of the old horse-and-buggy type, even if they held no degrees. They were men who, in emergencies, could work miracles of medicine and surgery. To this school of sturdy disciples of Hippocrates belonged Pharmacist's Mate M. C. Watts of the S. P. Lee. Like most of his shipmates, who had gone to their bunks before the destroyer hit, Watts had rushed on deck without taking time out to put on shoes or add to his apparel beyond the skivvy shirt and shorts usually worn aboard ship in lieu of pajamas.

At the base of the cliff, Watts had seen the wounds inflicted upon his comrades as they climbed upward. Mustering his courage—which was the fruit of the sheer self-sacrificing determination that is the hallmark of his profession—the S. P. Lee's only medical man scaled the wall as hurriedly as he could without regard to the cuts and bruises inflicted upon himself in the process of his breathless climb. When he reached the mesa—and after paying but cursory attention to his own wounds—Watts set about to minister to the injuries of his shipmates.

He wiped their cuts as best he could and, to stop bleeding, bound them with rags torn from garments that they could ill afford to spare, or from the cotton fabric of life preservers. Watts was hard at work—and wondering how he would be able to care for his steadily increasing number of patients—when lights like those of lanterns were observed to emerge out of the fog from what appeared to be landward. The new arrivals proved to be John Giorvas and his section hands.

At this particular time, the few officers and chiefs who had gone over the side of the destroyer were still down below urging the men on the ledge to tackle that last perilous obstacle—the cliffs of Honda—to save themselves and to make room on the narrow shelf for newly arriving shipmates. Thus it was that the role of giving the outside world the first actual news of the disaster fell to Pharmacist's Mate M. C. Watts. He told Giorvas all he knew—that three destroyers were on the rocks for sure—the S. P. Lee, the Nicholas, and the Delphy. He added, urgently:

"We'll need doctors and medical supplies. Some of our men are really badly cut up. I've no idea what conditions are on the other ships, but it could be that many of their people have been seriously hurt. Please get as many doctors and medical supplies as you can—and pronto."

Giorvas told Watts about the not-so-distant Section House and explained that it would take time to bring help to Honda. He left his laborers to give whatever aid they could to the men of the S. P. Lee and headed, on a half-run, back to the Section House to telephone General Foreman Maes.

Meanwhile, at the end of a long, tough hour, the last two men to leave the stranded destroyer reached the top of the mesa. They were Captain Morris and Commander Toaz. By then, nearly all of the badly injured men from the S. P. Lee had been led or carried, by their more fortunate buddies or the section hands, to the shelter of the trees along the railroad tracks at the site of the Section House. There, bonfires fed by railroad ties soon were roaring. Captain Morris remained on the top of the cliff to supervise the setting of the signal watch for the Nicholas.

As for Commander Toaz, with two officers and about a dozen volunteers, he headed south along the rim on Honda mesa in search of the Delphy and to give such assistance as he could to Squadron Commodore Watson and Captain "Dolly" Hunter.

Riding the Rails to Honda

On the station platform at Surf, City Marshal Bland and Editor Adam were met by General Foreman Maes and a couple of railroad hands. The fog was so thick that the motor track car and its crew were barely visible on the rails a few feet away.

"Are you coming with us?" asked the Marshal.

"No," replied Maes, "I'm trying to round up a couple of doctors over in Lompoc and I'll bring them up myself as soon as I can get hold of them and get them over here."

"Many people hurt?" asked Ronald Adam.

"Don't know for sure. I can't make it out. Either Johnny Giorvas, my Section Foreman at Honda, has gone loco or all hell's a-popping up there on a grand scale. Johnny says that three destroyers are aground that he knows of and there may be more. So I thought that you'd best get up there as fast as you can, Marshal, to see what's what."

"Okay," snapped Bland, who was a man of action. "Let's get going!"

The motor on the trackcar spluttered into life.

After some nursing by one of the railroaders, it settled into an irregular cough and the vehicle rolled off over the slowly rising 5½-mile grade toward Honda.

Contrary to the contention that motors run best in moist weather, this engine acted badly all the way. The railroad men blamed the trouble on the timer. At any rate, the car slowed down several times for lack of power. Twice it came to a dead stop as the engine died out. But each time, the SP men cranked the motor back to life. On the third stop, Bland's temper came close to the boiling point. The fog was thick. The night was cold. The rails were useless. He had a job to do and he could not get a chance to do it. The lawman watched the vague shapes of the railroaders as they bent over the track car's engine. Suddenly one of them jumped as if he had been jolted by an electric shock.

"Quick! Everybody!" he shouted. "Get the car off the track. There's a train coming."

In desperate haste, with everybody lending a hand, the car was lifted off the rails. They had barely placed it beyond the line of ties when another voice yelled: "Down! Quick! Down or the cars will suck you in and kill you."

He had not finished his warning before a huge blur of light from the beam of a locomotive headlight exploded out of the fog from the direction of Honda. As it descended upon them like a living, malevolent force, they heard the rumbling exhaust of the locomotive and the thunderous roar of swiftly rolling wheels.

One moment the train was there, bearing down on them like a vengeful juggernaut. Then a wild burst of howling air; a cyclone of cinders and sand; the bang of wheels hitting the ends of rails; invisible fingers of rushing air dragging at their clothing and bent on pulling them under the wheels. Then, just as suddenly, the roar of the train receded and it was dark and silent again. But the darkness and the silence and the fog were more frightening than they had been before.

"That was a narrow shave," observed Adam to the man who had shouted the warning. "We might have had a major railroad wreck on our hands, had we lived to see it. How did you know that a train was coming? In this fog we could neither hear it nor see it!"

"I felt it," replied the railroader. "I felt the vibration of the rail through the soles of my feet."

When the party reached Honda Station, it was greeted by the flares of rockets sent aloft from off Saddle Rock. (These were probably signals fired by the Nicholas to enable the signal watch on Point Pedernales to establish her position.)

After crossing Honda Canyon trestle, the track-car was abandoned. Bland, Adam, and their party made their way down the wagon road that led to the Saddle Rock quarry. Oddly enough, they did not run into Lieutenant Huff of the S. P. Lee and his signal watch. But that could have been because the Lompoc men approached Point Pedernales on the north side of Saddle Rock. Huff and his watch could have sought shelter on the lee side of the outcropping and still have kept watchful eyes on the Nicholas.

As Ronald Adam reported: "Although the fog was extremely thick, the rockets that were sent up made it possible for us to distinguish the outline of a vessel close inshore. The ship that we could see was the S. P. Lee, #310, and we did not learn until later that another vessel farther out than the Lee was the one that was sending up the distress signals."

At that time, the members of the relief party had no way of knowing that the S. P. Lee had been abandoned and its complement taken to Honda Section House. "A consultation was held and we decided that the best aid we could render would be the building of a big bonfire on shore and having our coil of cable on hand if it were needed. No wood was to be had in this bleak region and some of the party started back to Honda Section House to procure wood and kerosene for building the fire." On the way to the Section House, the party encountered Captain Robert Morris (ComDesDiv 33), who was wandering in a bewildered state on the sandy mesa that separates Honda from the ocean. "It was from Captain Morris that we learned that there were several vessels in distress and that all of his men had been landed from his flagship. He did not know where his men had disappeared to."

Around the Section House, the fog lay so thick that Captain Morris did not see the flames of the huge bonfire near the track until close upon it. Built of discarded railroad ties, which were stacked like cordwood and in good supply, the fire roared hot and lustily. It gave a welcome and cheer-inducing warmth to the scores of men of the S. P. Lee who were clustered about it.

By the light of the flames, Captain Morris, for the first time, was able to see that his men were not only a bedraggled-looking lot but that many were suffering from bruises as well as badly bleeding cuts. Reported the S. P. Lee's Pharmacist's Mate Watts to Captain Morris: "Sir, I have talked with Section Foreman Maes over the telephone. He's at a station up the line called Surf. He is coming up with two doctors from Lompoc who have volunteered their services."

"Good work," snapped Morris. "Where is the telephone?"

On being taken to the SP instrument in the Section House, Captain Morris contacted Maes and explained that, in all probability, several hundred wet and hungry men—many of them only half-dressed—would need not only medical attention but also clothing, blankets—anything to keep them warm. Hot coffee and something to eat was also an urgent need.

"Several hundred men?" gasped Maes as he realized the extent of the disaster. Then he squared his shoulders: "It's a large order and Honda is a helluva isolated spot—you could not have picked a worse one—but, Captain, you can bet your bottom dollar that we'll fill it."

After hanging up, Walt Maes got busy on the telephone to Lompoc—clothing and food were needed at Honda. Next he talked with Trainmaster Foley at San Luis Obispo over the telegraph wire and stated not only the needs but also added that at least three destroyers were aground at Honda. Foley promised immediate action. But first he passed the new information on to Night Chief Dispatcher Keller in San Francisco. This time the story of the stranding was not disregarded. The newspapers were notified. Soon the city rooms of the Chronicle and the Examiner were stirred out of the usual Saturday night doldrums. The Sunday mail editions had gone to bed. But there was time for make-over on the city editions. Telephone wires began to hum with activity; typewriters and linotypes clattered, and the wires and cables of news services were fed a fast flow of important tidings. The trouble was that, as incomplete or exaggerated details of the disaster flowed eastward across the country in the news streams, the story entered into time zones that brought it to editorial attention at later and later Sunday morning hours. Way back East, for instance, nearly all the Sunday newspapers were already off the press and in the lanes of circulation when it broke. Few cities have Sunday afternoon newspapers, and radio, in 1923, had very incomplete news distribution.

This meant that information about the disaster was very spotty across the nation on that Sunday of September 9. This lack of accurate information and a superfluity of wild reports were to play vital and vicious roles in the crop of rumors, recriminations, and false statements which always fill a news vacuum in any controversial matter of overwhelmingly great public interest. And that, to be sure, was just what happened in the first telling of the story of the Tragedy at Honda.

Death of the Delphy

George E. Jordan, Pharmacist's Mate of the Delphy, was sitting on the edge of his bunk. He had stowed away his clothing and had just removed his shoes preparatory to turning in when, as he later expressed it, "the bottom seemed to fall out of everything." Following a tooth-shattering jar, the men in the forward crew space scattered around the deck like tenpins struck by the ball of an expert bowler. Next, the destroyer pitched and tossed so wildly that Jordan and his companions were rolled from port to starboard in a tangled heap. Human cries and curses mingled with the shrieks, moans and splintering sounds of the Delphy's tortured steel.

"I headed for topside as soon as I could stagger to my feet," explained Jordan. "Getting up the ladder was not easy. The old girl smashed back and forth and breaker after breaker dumped tons of sea water down the open hatchway. Then the lights went out. By the time I put foot on the deck, I was as briny as a herring but I also had the night vision of an owl.

"Even that being the case, I could hardly believe what I saw as I peered off the destroyer's starboard quarter. It seemed as if other destroyers, like gigantic meteors, came streaking out of the fog and shot down upon us to port, to starboard, and astern. They came as fast as greased lightning, like death at twenty knots. Then it came to me that I might have a job on my hands, and that I would not be at my best without clothing and footgear. So I made my way below as best I could. It was hard going bucking my way down the ladder with dozens of men swarming up it."

Lieutenant Richard Cruzen, on making certain that all precautions had been taken in engine and firerooms, ran on deck and down the port side to see what had happened. There had been no collision, such as he had anticipated, but from the fantail he could see rocks to starboard.

"I then realized," he said afterward, "that we were aground rather than in collision. As I turned my gaze from the rocks and looked astern, some ship flashed on its searchlight. By its beam, made somewhat blurry by the fog, I saw a destroyer, the Young, lying on her beam ends. As far as I could see, no men were clinging to her side. Believe me, it was a very distressing sight, especially as I was very fond of the Young's skipper, Commander Bill Calhoun."

But, distressing as it was, young Cruzen had to turn his thoughts from the destruction of the Young to the immediate demands of the Delphy, which was also in mortal peril. She was rolling heavily in the surf with her starboard side to the cliffs. A rock had opened the after-oil tank and was splitting a huge rent in the deck plating. Making his way forward, Lieutenant Cruzen found the majority of the crew assembled in the well deck. They were wet, cold, and miserable, but there was no sign of panic or confusion.

Just then Coxswain E. B. Palmer reported that the starboard whaleboat had broken loose. With the aid of a couple of men, they shortly got the boat, which was swinging dangerously over the deck, secured. About this time, Lieutenant Cruzen noted that clusters of men were moving aft from the well deck. Concluding that the order had been given to abandon ship—as yet to be issued by Captain Hunter, who was still on the bridge—Lieutenant Cruzen informed Commander Harry Donald that he knew a way to get the men off the ship. He told of his visit to the fan-tail and of having seen a large rock right under the destroyer's stern. With Donald and a group of men, Cruzen hurried aft.

At first, the stern was close enough to permit some men to jump to a lava rock from the fan-tail and the propeller guard. Several of the crew made the most of this opportunity to leap to safety. But this easy avenue of escape was soon to be closed. The surf, roaring in—high, wide, and ruthless—rolled the ship violently. This, in turn, opened up her bottom, and the distance between the stern and the rock became too great to jump.

At this juncture Captain Hunter, realizing that the situation was hopeless, passed the word to "Abandon ship" and started aft to the fan-tail where "Blinky" Donald and Lieutenant Cruzen were already attempting to get a line to a rock about 20 feet from the starboard quarter.

As he passed the starboard sea ladder (steel rungs welded onto the ship's side down to the waterline), he saw several men in the water with Raymond L. Rhodehamel, Engineman 1st class, attempting to rescue them. Three of the men had evidently jumped overboard from the bow attempting to make the shore—actually the northern side of Bridge Rock—but immediately found themselves overwhelmed by the crashing power of the seas that swept in between the Delphy and the rock with the fury of a tidal wave.

Fortunately for the would-be swimmers, the water near the ship was just then being covered with a rapidly thickening coat of fuel oil from the punctured forward oil tanks. While it was deadly stuff to swim through, the blanket of grease did tend to reduce the turmoil of the sea and enabled them to reach the relative safety of the Delphy s side right under the starboard sea ladder.

Rhodehamel was making his way forward to the life-preserver locker when he heard their cries. Without stopping to consider the risk he took, Ray descended to the bottom rung of the sea ladder and, lowering himself from it, performed the truly herculean task of snatching his shipmates from the greedy sea.

Ensign Morrow, who happened along just at the instant Rhodehamel went over the side, also went down the ladder and lent the daring engineman a much-needed helping hand. Later, in recommending Rhodehamel—with Bayes, Roseboro, and Etheridge—for the Life Saving Medal, Captain Hunter stated: "... although the violent plunging of the ship made his position hazardous to the extreme, Rhodehamel succeeded, through great effort, in rescuing the men in the water, and getting them on the deck of the ship. What with the darkness, the slippery condition of the oil-covered sea ladder, and the insecure footing thereon, and the ceaseless lurching of the ship, Rhodehamel was in constant danger of losing his life. And yet, he stuck to his task until all of the men in the water were recovered by him."

Last to be rescued by Rhodehamel was James T. Pearson, Fireman 1st class. Small of stature but great of heart, Pearson had actually jumped overboard to help save his three floundering shipmates. But when he hit the water in a hard belly landing, Pearson's glasses were broken and pieces of glass became embedded in his eyes. Blinded, almost mad with pain, Jim threshed about in the sea. Twice he vanished from sight. Undoubtedly he must have swallowed and inhaled a dangerous amount of fuel oil.

Quickly, Morrow passed Ray a line. Again showing the heroic stuff he was made of, the engineman let go his hold on the sea ladder. After a desperate struggle, he succeeded in getting the line under Pearson's arms. As the pain-crazed fireman struggled with insane fury, the pair were hauled back to the sea ladder. Rhodehamel made the bottom rung. But when it came Pearson's turn to be pulled aboard, the situation changed. Now began a desperate and dramatic rescue effort in which the Delphy joined the sea in playing the role of villain.

Each time the ship rolled away from Pearson, the line would be stretched tight. But as the vessel snapped into a reverse roll, it was impossible to take in the slack fast enough. To Morrow, it seemed almost inevitable that the fireman would be crushed between the ship and pinnacle rocks along her starboard side. As a last desperate effort, and at the risk of suffocating Pearson with fuel oil, the line was moved aft. The unfortunate man then was towed to a position abaft the point where the ship was rolling against the outlying rock. There, with the aid of men on the starboard propeller guard, Rhodehamel was able to get his man aboard.

Lifting the now unconscious man in his great arms, his rescuer carried Pearson forward and laid him, face down, across a chest where he was protected from the cascading seas. Pharmacist's Mate Jordan came on the run to help revive him. Suddenly, before resuscitation could be begun, Pearson leaped to his feet, wrenching himself from Rhodehamel's grip with the strength of a maniac, crazed by the unbearable pain in his eyes and by the effects of the fuel oil. He plunged about the deck, slamming blindly into obstacles, while his shipmates tried to control him.

Just as it seemed he would go overboard, Jim Pearson toppled, cut and bleeding, upon the steel deck. At this point, Lieutenant Blodgett, the Exec, took charge. While Pearson was unconscious, Jordan—with the aid of flashlights held by Blodgett and Morrow—tried to remove the glass splinters from the fireman's eyes. However, they were too fine to be seen and Jordan lacked proper instruments. Through almost superhuman efforts led by H. J. Fontan, Gunner's Mate 1st class, a life raft was actually launched. But before Jim Pearson could be placed aboard it, the craft was sucked away from the ship's side by the action of a receding wave and dragged out to sea. Fontan, the only man aboard the raft, had just time to make a frantic but successful grab for the sea ladder, before the Carley headed out for the fog-shrouded ocean.

Later, Blodgett, despite his own injured knee, tried to get Pearson on a life raft. But he was so slippery with oil and struggled so violently that nothing could be done with him. It was unsafe to put Pearson on a raft in the state he was in. He would be a serious menace to the safety of anyone who might be aboard the raft with him. To ensure Pearson's safety until he could be evacuated, Lieutenant Blodgett had E. R. Wiesendanger, Radioman 3rd class, loop a signal halyard under his arms and lash him to an angle iron that supported the destroyer's searchlight tower. That was done so that Pearson would not roll over the side. If he should recover his senses, Jim could easily free himself by slipping his arms out of the loop.

The actions just described took place much more rapidly than they can be told and the situation of the Delphy worsened just as rapidly. By the time Hunter reached the fan-tail, heavy seas were breaking over the ship forward, preventing anyone from attempting to save records or valuables from the ship's office or wardroom country. A line to the rocks, for getting men ashore, had to be rigged immediately if heavy loss of life was to be prevented.

Leaving Pearson as well provided for as was possible until he could be rescued later, Lieutenant Blodgett hobbled painfully aft looking for a solution to the evacuation problem. It appeared to him that a strong swimmer could make it to the first rock some 20 feet from the ship's side.

Wiesendanger immediately volunteered. "I can do it, Mr. Blodgett. I'm a good swimmer, let me try!"

"All right," said the Exec "With your name and your guts, I think you can do it. Good luck; you'll need that, too."

A line was put around his shoulders and Wiesendanger made his way down the slippery rungs of the sea ladder.

Once there, Wiesendanger, apparently concluding that the line would encumber him, cast it from him, dove into the sea, and set out for the rock with powerful overhand strokes. Still, strong swimmer though he was, the radioman faced serious trouble. Twice, he was washed completely out of sight beneath the Delphy's bottom by the vicious undertow. Twice, miraculously, he escaped being crushed to a pulp by the rolling hull of the ship. But eventually, almost completely spent, Wiesendanger reached the rock and laboriously crawled up its steep and oil-slicked side. A great cheer rose from all on board the Delphy as the heroic swimmer staggered to his feet and signaled for someone to heave him a line.

Meanwhile, H. H. Wilgus, Boatswain's Mate 2nd class, had broken out from the searchlight structure a mooring line to be used as a life line from the doomed ship. Rigging this line from the after deckhouse to the rock was extremely hazardous. On two occasions, while attempting to get this job done, Wilgus was washed overboard from the propeller guards by one sea, only to be slammed back by another. Although his leg had been badly injured, Wilgus stuck to the job until it was finished. In this he was materially aided by Chief Signalman H. D. Cummings, T. J. O'Hare, Quartermaster 1st class, and G. H. Connon, Machinist's Mate 1st class. Despite his injuries and numerous bruises, the boatswain's mate refused to seek safety ashore but remained on the deckhouse assisting the other members of the crew to the shore after the two lines of passage had been swung.

Once the mooring line's outboard end had reached the rock, Wiesendanger took a couple of turns around a fairly smooth bollard-like piece of rock. In that manner, he would be able to keep the line fairly taut for those who were to follow him, and, at the same time, give and take whatever amount of line was necessary to follow the swaying motion of the ship.

First to test the security of Wiesendanger's somewhat precarious life line was Lieutenant Commander Donald. With the air of indifference of a man who had learned to take life in his stride, "Blinky" took the line in a firm grip and reached the rock without incident. He was quickly followed by Lieutenant Cruzen and W. J. Ekenberg, Seaman 2nd class.

The latter carried a length of boat falls—a line used to hoist the boats to the davit heads or to lower the boats into the sea—which he had volunteered to take from the first rock to the bluff face of Bridge Rock. On reaching his first objective, Ekenberg did not take time out to ponder about the how's of the disheartening layout of surging surf and slippery, knife-edged lava that awaited him. He plunged through the water and, despite the terrific drag of the racing undertow, he managed to climb above the reach of the oil-coated waters and found a place to secure his end of the boat falls.

Meanwhile, on the rock Wiesendanger and "Blinky" Donald had made the other end of the second line fast. Again Commander Donald served as guinea pig to test a life line. Hand over hand, he swung toward Bridge Rock's northern bluff. The fact that he made it proved the line secure. From the bluff, Donald again checked to see that the line was well secured and then signaled okay to Lieutenant Cruzen, who would be in charge on the rock.

Cruzen, in turn, passed the okay to Ensign Morrow, who directed the abandon ship operation on the after deckhouse. Men were needed on the first rock and on the bluff to receive and help their shipmates as they came across the lines. Volunteers for these hazardous jobs, which offered painful and uncertain footings and constant dowsings, were not slow in stepping forward. Selected for the rock were Chief Carpenter's Mate J. G. Whittaker; Chief Yeoman J. C. Holdorf; Coxswain E. B. Palmer; G. J. Burke, Yeoman 1st class, and L. G. Robbins, Fireman 2nd class. To assist Lieutenant Commander Donald and Seaman Ekenberg in hauling the men up to the top as they completed the second leg of their hand-over-hand trip, the following were selected for the bluff: Chief Torpedoman H. B. Wilson and Coxswain H. A. Isbell.

The reason for the larger number of men on the rock was that, as the ship rolled, the men on the rock would have to haul in on the line or pay out in order to keep it taut despite the never-ending lurching of the Delphy. Up to now, the transfers had been working smoothly; in fact, things had been going with deceptive ease. When James W. H. Conway, Fireman 3rd class, ventured out on the life line, affairs took a disastrous turn.

Lifted by a monstrous breaker, the Delphy careened so far to starboard that, despite the efforts of his shipmates on the rock to keep it taut, the line slackened, Conway was dropped into the seething waters, and his legs were caught against the rock in a bone-crushing vise.

Despite the agonizing pain, he clung for dear life to the now slack life line. Quick as a flash, the outgoing ground swell threw the Delphy to port. Before the men on the rock had time to pay out line to meet this situation, the line snapped as taut as a piano wire.

Acting like a huge sling, the mooring line flung Conway aloft, breaking his hold on it. The unfortunate man plunged downward and his horrified shipmates heard his piercing scream as they saw him fall into the sea. Although gallant efforts were made to recover him, as he was carried past the propeller guard under the fan-tail, Conway's clothing was so covered with fuel oil that James Farrell, Gunner's Mate 1st class, who attempted the rescue at great risk to his own life, could not maintain his hold upon the slippery garments.

Fireman 3rd class T. C. Farnham also had a horror-filled experience in trying to reach the first rock, but his luck was better than poor Conway's.

His struggle with death began when he lost his grip on the oil-soaked mooring line and plunged into the raging maelstrom of fuel oil and salt water off the Delphy's side. His chances of survival would have been zero but for the quick action of three men. Acting as one, they dived into the sea between the ship and the rock.

It was hazardous work and succeeded only because the three men: A. A. Bayes, the Ship's Cook, together with J. C. Roseboro, Gunner's Mate 3rd class, and W. C. Etheridge, Fireman 3rd class, acted quickly and efficiently with complete disregard for their own lives.

Down through the ages of history, men of the sea have had a flair for laughing at danger and making light of death. Many are fatalists; many are proud to repeat the old saying: "My ship is my coffin, my grave is the sea."

And, like the real sailors they were, the men of the Delphy: were they green gobs just out of boot camp or shellback veterans: responded instinctively to the traditions of the Service and to the "esprit de corps" that is engendered by Navy discipline. There was no hysteria. No crowding. The crew obeyed orders as they were issued and calmly waited their turn.

Still, the death of Conway cast a somber pall across the steadily diminishing group as man after man stepped up to the jumping-off point where Chief Signalman Cummings placed a heaving line around his waist and started him shoreward, hand-over-hand, on the mooring line.

Chief Cummings' brisk, matter-of-fact, yet jocular air had such a calming influence on all concerned that Captain Hunter recommended him, as well as Wiesendanger and Ekenberg, for the Life Saving Medal. About 2200 hours, Lieutenant Blodgett reported to Captain Watson and Commander Hunter that all hands had been disembarked except James Pearson. When conscious, he was unmanageable and raving with uncontrollable fury. When unconscious, he was a dead weight too heavy to carry under existing conditions.

There was just no way of getting the injured man safely ashore. Because of his violence, he could not be carried or hauled across the lines. The blanket of oil between ship and shore was now so thick that to venture into it could easily mean death. And the sad fate of Cabin Cook Sorfornia Dahlida—who was choked and drowned by oil that filled his throat and lungs when he slipped from the life line and fell into the sea—proved that to try to swim Pearson across would be utterly impossible. To leave Pearson alone, lashed to the searchlight tower on the deserted ship, was a hard decision to make. But there was no other course to take. With daylight and better sea conditions, it would be possible to take Pearson ashore via a breeches buoy or in a life raft.

To that possibility they must pin their hopes. So far as the nature of its terrain is concerned, Bridge Rock is fashioned from the same sharp and cutting lava that the face of Honda is made of. The only difference is that where one is almost vertical, the other is fairly horizontal. Although better, at any time, than a sinking destroyer, Bridge Rock is no bed of roses, as its unwilling guests were soon to discover. Like those aboard the S. P. Lee, many of the Delphy's men were in their bunks when the crash came. And since the inrush of the sea and the order to abandon ship came so swiftly, few, if any, had any time to put on shoes and clothing. After a few steps, following their landing on the bluff, the feet of the men were terribly cut. Commander Donald solved this problem by setting men who wore shoes to carry their shipmates from the landing level to higher ground. Before he came ashore, Captain Watson had issued orders for all personnel to remain at the place of their landing.

At the time, he had in mind the uncertainty of their position as well as the possibility that the sea might quiet down to a point where it would be possible to carry on salvage operations. Busiest man on Bridge Rock was Pharmacist's Mate Jordan. Like his opposite number on the S. P. Lee, George had a large number of cut hands and feet to attend to. But, more serious than that, a full dozen men had received major contusions and bruises. Fully half of these had suffered severe shock.

So far as Jordan could see, some of his shipmates who had escaped death on the Delphy might possibly die on this miserable rock unless they received medical care promptly. He knew that the Division Medical Officer rode on the Woodbury—but where was the Woodbury? Scuttlebutt was that no one really knew where they were but that the best bet was that they were stranded on an isolated rock off the Santa Barbara Channel. Then three long, lonesome calls of a locomotive whistle reached the ears of the men on the rock.

Some swore that they could actually hear the rush and rattle of the train. This truly momentous happening, which proved that they were not castaways on a desolate island far from any possibility of immediate rescue, acted as a tonic to all hands. Ensign Morrow immediately started exploring with his flashlight and presently discovered a path—worn, no doubt, by the feet of abalone seekers—that led toward higher ground, actually to the cliff of Point Pedernales itself at the edge of Honda mesa. That was enough of a lead for the venturesome Ensign Morrow. With the Commodore's permission, he and several others who had shoes, climbed a steep and rocky slope and set out through the low brush toward the railroad track. Shortly thereafter they saw a light which led them to a group of sailors and section hands ringed around a fire of railroad ties. They were the men from the S. P. Lee. And they kept on drifting in as Morrow watched. Black as imps from Hell.

Shoeless, in many cases. They limped across the rough ground; some carried their shipmates. Few wore even a complete covering of clothing, let alone foul-weather garments. They were a drenched and shivering lot, but the sight of that fire put new heart into them. Soon they had its flames roaring high as they warmed front-sides and backsides, dried wet clothing, wise-cracked, and sang songs; irrepressible even in disaster. And many a man silently thanked God for his deliverance.

"The next day," noted Morrow, "I saw that what had seemed a path was actually a narrow natural bridge. It was only about 3 feet wide, 25 feet long and about 30 feet above a swirling mael-strom. Using flashlights and on such a dark night, wherein one moved by feel rather than by sight, it did appear to be a path. That not a single soul—who afterwards walked this slim thread to safety that night—plunged into the watery abyss was really a miracle."

In passing, let us note that there were to be no salvage oper-ations aboard the Delphy. In the gray light of the early dawn, it was plain that there was no way of reaching her. She had broken in two and 30 feet separated her stern and forward sections. Eve-rything else was unchanged.

Stout Hearts Aboard the Young

The abandoned lifeline hung like a limp garland from the af-ter deck-house; seas swept over her deck with even greater force than they had during the night, and a blanket of fuel oil covered the waters between ship and shore. Of James T. Pearson, Fire-man 1st class, who had jumped into the sea in the hope of saving comrades who were about to drown, there was no sign. When Captain Hunter, with other officers and crew returned to the Del-phy at daybreak for the purpose of saving Pearson, they, to quote "Dolly" Hunter, "found no trace of him and there was no way of getting aboard. While we were watching, the pounding of the seas carried away the starboard supports of the searchlight tower. As this structure listed more and more to port it carried the ship over on her beam ends where the seas filled everything."

At the instant hidden rocks ripped her thin hull plating along almost the full length of her starboard side, the Young entered upon her death throes. From bow to stern, watertight compart-ments with bulkhead doors securely dogged, were instantly flooded. Under the pressure of inrushing seas, entrapped air, and the wrenching gyrations of the ship, bulkheads bulged, seams ripped, and rivets popped here and there. Air, followed by spurting streams of water, whistled out of the leaks.

The furious battle between the suction of the sea and the entrapped air in the ship caused a mastodonic rumbling in the depths of the stricken destroyer. A hideous and frightening sound that added a sharper note of fear to the terrors of shipwreck at night. There is nothing more demoralizing to coolness and discipline among men of the sea than to feel their vessel sink beneath them. And there, the Young was not only settling in the water but also, second by second, taking a dangerously heavy list to starboard. In 90 seconds, the destroyer was flat on her side with her port quarter only 18 inches out of the water.

As she turned, most of her gallant destroyermen, in the instinct of self-preservation and in response to shouted orders, followed her over until they gained a perilous refuge on her side. Those of her crew who, for some reason or other, did not follow the Young in turning were swept into the sea and either drowned or met their death on the sharp teeth in the Devil's Jaw.

That the Young lost many of her crew was not due to any breakdown on the part of the life lines of discipline. With but few exceptions, every man who was on deck at the time of the disaster saved his life by obeying Captain Calhoun's order to climb on the port side. Those who were reported as missing were mainly members of the engine room and fire-room crews on duty at the time of the stranding, or they were men sucked away by the undertow as they poured out of the hatches from the crew's spaces while the ship rolled over. But, to the men who clung to her upturned side, their position was precarious in the extreme as the Young, with convulsive shudders, sank deeper and deeper into the raging black waters.

At the engine room hatch, it did not take "Red" Hall long to convince Chief Kerrigan that the Young was in serious trouble. He and Hall rushed into the galley deckhouse passageway to join the crowd of men who were grabbing life preservers in the overhead. Kerrigan, to make room for others, went out on deck. He was just in the act of tying his life jacket when he found himself dumped into the sea by the sudden list of the ship. He spat out a slug of salt water, but not his fresh cud of chewing tobacco.

"Why the hell," thought Kerrigan, remembering the premonition he had had about this trip, and the hunch of the Chief Commissary Steward. "Why the hell didn't I 'jump ship' like he did?"

The crashing of a huge comber took his thoughts away from this unproductive line of reasoning. He would have to do some fast thinking to get out of this mess alive. As luck would have it, Kerrigan did not lose the waterproof flashlight he always carried in his hip pocket.

As he rose and fell on the heaving seas, he saw, by the flashlight's beam, men massed on the highest part of the hull as well as some clinging to the mast that extended out to starboard. But, in the tumult, no one paid any attention to Kerrigan. "Dear God," he prayed, "help me now. I've got to get home. You just can't let me drown now."

As if in answer to his prayer, a deeper shadow broke away from the shadows huddled on the deck. It proved to be Radioman Reddock, who had been distributing life preservers. Holding onto one corner of a life jacket, he cast the other to the Chief Electrician. Two other men joined to help drag him aboard, and, presto, Kerrigan was once more on the Young—although minus a lot of skin off his right hip.

"Seems like you're being baptized instead of your baby!" cracked Reddock. "Or is it that you need a diaper?"

Kerrigan grinned and turned to lend a hand in the distribution of life preservers.

A few minutes earlier, Lieutenant (jg) Felix L. Baker, Communications Officer of the Young, had been en route from the wardroom to the radio shack when the destroyer was raked by the rocks. As he entered the radio room, Clitus Reddock, Radioman 1st class, turned away from the speaking tube to the bridge. "Sir," he said to Lieutenant Baker, "I was just calling the bridge. Should I send an SOS?"

"No, boy," replied Baker, "you get the hell out of here—but fast!"

Just then the speaking tube to the bridge buzzed its peremptory summons. Reddock answered, listened, said: "Aye, aye, sir!" and snapped the cover shut. "That was the Captain," he explained. "He told me to get up on port side and follow the ship, if she should turn over."

"Okay, Sparks, get on the high side and make it snappy," answered Baker, and dashed for the bridge. As he reached it, he saw Coxswain Charles A. Salzer desperately trying to get men—who had taken refuge on the now horizontal foremast, its stays and ladder—to abandon their places of poor security and obey Captain Calhoun's order to stay with the ship.

After some shouting, they came—all but one man who insisted on clinging to the yardarm. He, too, at last was talked to salvation. The list of casualties, compiled the following morning at roll call on Honda mesa, reveals that the Young's engineers and firemen held their battle stations at throttles, gauges, and oil burners even as the destroyer careened and took them to their death. Among those who made it to the deck from below too late to reach safety were one engineman and seven firemen. One of the most courageous of the many brave men aboard the Young that night was a freckle-faced, snub-nosed fox terrier of a man, Fireman First Class F. T. Scott. The Young was listing dangerously; her engine-rooms were flooded; her Engineering Officer, Lieutenant (jg) Frederick C. Sasche, had been dragged on deck, unconscious from the fumes below. The danger was that, with the oil burners still pouring heat into the boilers, and the turbines cut off, steam pressure might mount with terrific speed and a boiler explosion tear the ship apart. Of this Captain Calhoun and his Exec, Lieutenant Herzinger, were keenly aware. They called for an engineer to go below. Those burners must be turned off—but right now.

"Let me go, Captain," shouted Scotty above the tumult; "I know just where the master valves are!"

The fact that these valves were below the floor-plates in a flooding fireroom apparently meant nothing to him. "All right, Scott, report back to me as soon as possible," shouted Bill Calhoun, "—and don't take too many chances."

Scotty disappeared down the fireroom air-lock; the Young rolled flat on her starboard side and all hands clambered over the rail onto the only space left above water. The boilers did not explode, but Captain Calhoun and his Exec, waited in vain for Scotty to report back for a richly deserved "Well done." After the destroyer turned over, "waves broke continually over her," according to Captain Calhoun's report. "The side became covered with fuel oil and was rendered very slippery and difficult of footing. The position of the survivors of the crew was extremely dangerous; prompt action was urgent and necessary."

Urgent and necessary. How great can urgency be? How strong the drive of necessity? The fourscore survivors of the Young's company clung to the sea-swept and wind-blown side of the destroyer in a space perhaps 6 to 8 feet wide by perhaps 25 feet in length on the oil-slicked port side of the ship just abaff her numerals—312—on the bow. Their only hand holds were the portholes smashed in by Peterson's axe. Working fast, at Calhoun's direction, Chief Boatswain's Mate Petersen set about recovering the hemp lines that are laced between the life lines of the forecastle. They form a sort of net to prevent personnel from being swept overboard between them. These hemp lines he unrove and, with the help of Herzinger and others, used them to bind the crew together—like so many stalks of asparagus—in groups of six or eight to give them added security. Arthur Petersen, "Pete" or "Ole" to his shipmates—and even "Big Pete" in some accounts—was a man of parts, a man to tie to. He was not a big man physically, perhaps 5 feet 10 inches, thin, rawboned, and wind-burned. He ran a "taut ship" in his domain and had no problems of discipline that he could not handle himself. Lieutenant Herzinger had tremendous confidence in him.

Urgent and necessary

It is as simple as dialing a TV channel, sitting here visualizing the plight of these men through the calm eyes of hindsight.

But think of the stark, naked fear that must have torn at their hearts and minds as they sat, squatted, or lay on that wave-swept and insecure platform of steel. All about them were the realities of an inferno that would dwarf the imagination of a Dante or a DeMille, yet their faces and actions gave no sign of anything but courage and determination to survive.

Destroyers, many of them—division on division—had come tearing out of the night on each other's heels and shot off in every direction like stampeding elephants.

Barely visible, except for fog-dimmed portholes and running lights winking in the night, the ships plunged and wallowed—sirens shrieking, whistles blowing, smokestacks belching great clouds of black smoke as they tried to back away from destruction with white water boiling along their sides from desperately reversing propellers.

Searchlights probing briefly through the night; rockets rising through the fog; and carbide bombs, spurting tongues of flame, floating briefly on the sea.

Then: All darkness. All silence.

Four score men huddled closely together on a small roost of steel like seals on a rock, 18 inches above the surface of the sea. Fourscore men who wondered what had happened to their destroyer mates and what was about to happen to themselves. It took discipline—self-command of the highest caliber—for those men of the Young to stick it out.

In that grim, danger-packed moment, they demonstrated the courage shown by Queen Victoria's Royal Marines in the Birkenhead Drill when they stood in ranks, as if on inspection, on the deck of the transport Birkenhead in 1852 as that doomed vessel was swallowed by the sea.

This stirring display of cool discipline was immortalized by Kipling in a similar disaster in 1893 when, as a result of a mistake in signals during maneuvers, the HMS Victoria, a battleship, was rammed and sunk. Her "Jollies" (Marines)". . . stood and were still to the Birkenhead Drill and went down with their ship, standing in ranks.

There is an old Bull Halsey-like maxim in the Service that runs: "Do the best you can with what you've got, sailor." And, aboard the Young, there was immediate and urgent need for doing just that.

The trouble was that Captain Calhoun had virtually nothing "to do" with. The nearest point of land—if one could call it that—was an outcropping of large rocks about 100 yards to the east (Bridge Rock). Under the prevailing conditions, it would be next to impossible to swim that distance with a line.

But, since it was not utterly impossible, capable, tough Gene Herzinger proposed to attempt it. Gene, a "mustang," which is the Navy term for an officer who has come up from the ranks, began his career at sea by shipping as a fireman and grew up in a hard school.

A deep scar on his forehead was a souvenir of a battle ashore one night in Rio. He was a man not easily bluffed by any situation. On this night, Gene made his way aft with several men to help him and unrove the boat falls—2-inch lines each about 100 feet long—from the motor dory. These they brought forward.

Meanwhile, Petersen secured a life ring from the bridge and some lengths of light line with which he proposed to swim to the rocks.

Just then, the Chauncey rushed into the gap between the rocks and the Young and, as already described, ended up by coming so close that the Young's propellers ripped through her sides and drowned out all power and lights. In arranging this incident, Destiny acted in the Young's favor and gave her men a new lease on life.

When the Chauncey was finally hurled high upon a ledge, her stern projected about 25 yards from the rocks and thus shortened the Young's escape route to about 75 yards. Not only that, but the location of the Chauncey near the southwestern wall of Bridge Rock reduced and diverted the heavy surges of receding surf that had been sweeping the Young.

The Chauncey Aids the Young

Captain Richard H. Booth, of the Chauncey, with his destroyer in a precarious position on a submerged ledge on a lee shore, realized that he had now been entrusted with a grave double responsibility which could mean life or death to many men. His first obligation was to get his own men safely ashore. His second, to give those aboard the Young every possible opportunity for survival.

But before either duty could be carried out successfully, communication by lines or otherwise would have to be established between the Chauncey and the forbidding western cliffs of Bridge Rock on the port hand. The distance between the vessel and the rugged, pock-marked volcanic wall was not great. Only about 25 yards. But the surge of seas within this stretch was fast and violent. Finally, two lines were carried ashore under the direction of Lieutenant "Chink" Lee with the heroic assistance of George Bleam, Water Tender 1st class. Ensign William E. Stock and J. M. Reynolds, Boatswain's Mate 2nd class, were placed in charge of the two Carley rafts that were to be used as ferries.

Although the Chauncey was the tenth ship in column and the last to run aground, her position was so much more favorable than that of the Delphy that Lieutenant Lee was able to complete the disembarkation of personnel in almost the same length of time that it took Lieutenant Blodgett, his opposite number aboard the Delphy, to establish his two-stage lifeline to the austere northern shores of Bridge Rock.

An interesting on-the-spot description of this scene was given by Radioman Frederick Fish of the Chauncey who, before he enlisted in the Navy, had had some experience as a newspaperman:

Sliding down the line to the raft, I found it covered with oil and so slippery that it was with difficulty that I retained my hold on it. The icy waters came up to our waists and my chief concern was to keep a package of "Bull" and some matches dry. The rocks on the shore were covered with oil and the men sprang ashore, skidding and falling over the rocks.

An unconscious man from the Young was discovered being tossed against the rocks by the waves. He was so oil covered that it was with difficulty that we held on to him and dragged his body on a rock. No medical aid could be rendered him on this precarious perch, so six of us started to haul him to the top of the cliff. The wall (of Bridge Rock) was almost perpendicular, but somehow, in the dark, we half dragged, half carried him up that cliff, and I noted that the men were very careful not to scratch him on the jagged rocks. Reaching the top, we discovered men from the Delphy lying all about.

The flickering rays of small lights revealed scores crouching behind rocks in order to keep their wet, half-clothed bodies out of the cold wind. It was so dark that it was almost impossible to see how many were injured or unconscious, and men were walking about in a dreamlike daze, stumbling and falling, cutting their hands and bare feet on the jagged edges of the cliff. And seemed not to care. Some of them, with the monotony of automatons, kept repeating the names of buddies as they staggered around in the manner of men who had lost their sight.

Many of us were under the impression that we were on a small island and that we should have to remain there until morning. Someone heard a train whistle, and we were astounded to find that we were on the mainland. I tried to roll a cigarette, but my hands were so covered with fuel oil that I found it impossible. After a brisk rubbing on the seat of my trousers I managed to get them fairly dry and rolled several cigarettes for some of the lads who were not so fortunate as myself. The fog and darkness did not permit us to see how many ships were on the rocks, and our only method of ascertaining was by asking men what ship they were from.

Standing on a point, I looked down and saw the Delphy being pounded by heavy seas, her lights out except for a lone light in the chart house which the Chief Radioman had hooked to a battery. This light remained burning for hours. At the stern of the Delphy a little group of men were working heroically, getting the balance of their crew ashore. Every few minutes a giant wave of cold water would burst over them and as it receded there would follow a clanking of metal as objects rattled and banged about the Delphy's decks.

From midships of the Delphy, even topping the pounding of the waves and the roar of the wind, came the cries for help of an injured man who was lashed there. He would have perished if he had not been lashed down. He perished anyway, but it was through no fault of the crew, as they did everything humanly possible to get him off. His calls kept up through the night, and they still ring in my brain. To hear a fellow creature calling for help and not be able to relieve him is the crudest torture possible to man.

An officer (Lieutenant Blodgett) came stumbling along the rocks. I questioned him about the man tied on the Delphy. He said that the man was blinded and covered with fuel oil, and could not possibly be taken across on the line. He had tied him to one of the searchlight stanchions in the best place he could find. He said the man was safe from the sea and could be gotten off in the morning.

Meanwhile, aboard the Chauncey, operations had been begun to get the men off the Young. Under the direction of Ensign F. A. Packer, who earlier had made an unsuccessful try at heaving a life line to the stranded sistership, the Chaunceys whaleboat was lowered over the starboard side. But, even before the stoutly constructed craft reached the water, a bruiser of a sea knocked the whaleboat into splinters. Since the motor dory could not be launched successfully from the port side, the only course left was to launch a raft, but—without a line to haul it to the Young—the poser was how to get the raft there in the face of contrary wind and waves.

Suddenly Chief Boatswain's Mate "Mike" Fyre pointed the beam of his searchlight out over the white-crested seas and shouted:

"Look! There's someone out there! Holy Moses, it's a man swimming and he's headed for the Chauncey!"

Chief Fyre, with Seamen Lyle Perry and G. S. Button hot on his heels, ran toward the fan-tail.

"You," yelled Chief Fyre to Button, "grab some line and stand by to get him aboard." To Perry, he shouted: "Keep the beam of my flashlight in front of him—not on him, that could blind him—so he can use it as a guide. I'll jump in and give him a hand."

Pete Peterson Saves the Day

Aboard the Young, soon after the Chauncey had ground to a halt, Chief Peterson stretched the boat falls from the motor dory on the side of the ship. He began to unlay the triple strands of manila. This gave him three lengths of line of about 40 yards each instead of one. Knotted together, they gave Pete a total of more than 100 yards.

This was easily enough to span the 75 yards between the Young and the Chauncey. Again Chief Peterson dived over the Young's side. This time he made for the port bridge wing where he obtained a small doughnut lifebuoy. To this he fastened one end of his manila line and slipped the doughnut over one arm and his head. The bitter (or inboard) end of the line Herzinger tied to a stanchion, and then he stood by, with others, to pay it out as needed.

All was now ready for this desperate bid for survival. Peterson turned to Captain Calhoun with a grin that lit up his sober face: "Permission to go ashore, sir?"—the usual request on going over the gangway for liberty. Nobody could resist that infectious grin.

"Permission granted, Pete—and God go with you," said the skipper.

Peterson turned and felt his way carefully down the sloping side of the Young until the seas were waist-high. Then, flinging himself forward, he struck out on his 75 yard race against death in the surging, oil-covered breakers. Never were his strong arms and sturdy body more severely tried. From the Young, eyes and prayers followed the swimmer as long as the darkness and fog permitted.

All hands hoped, with all the power at their command, that Pete would be able to make it. The manner in which the coil of line, held by Lieutenant Herzinger, kept slowly paying out gave evidence that the Chief was making steady, if slow, progress toward his goal. At long last—it seemed like hours but it took only minutes, and almost all of the coil had been paid out—there came three firm yanks on the line. This was the prearranged signal that Peterson had landed and that the heavier boat falls from the dory were to be bent to the lighter line and hauled across.

This was done. As soon as the big line had been made fast at both ends, Lieutenant Herzinger volunteered to go over the line, hand over hand, to the Chauncey to help handle operations. He obtained Captain Calhoun's grave consent. Good Execs like Herzinger were hard to come by. On the Young, the skipper himself took charge of the disembarkation.

"Aboard the Chauncey, they got a seven-man life raft and we on the Young hauled it in," explained Captain Calhoun later. "Peterson had tied boat falls from the Chauncey to the stern of the raft and thus we had a ferry. He came back with the raft and we then started evacuation. The first boatload included Lieutenant Sasche, Chief Kerrigan, Peterson, and four men. The Exec, of the Chauncey and Dick Booth gave first aid to Sasche. The raft was hauled back and forth making a total of 11 trips, and it was 2330 before we got all hands on the Chauncey. Then, finally, I, with my officers, abandoned ship."

During the long-drawn-out evacuation of the Young, with extensive periods of waiting between each return of the ferry, Captain Calhoun—although he felt heavy-hearted—led his men in the singing of songs.

They ranged all the way from well-established Navy tunes to popular airs of the day. Most of the men sang lustily. A few sat in moody silence. When the last enlisted man had left, Captain Calhoun—carefully assisted by Lieutenant Herzinger, who had returned to the Young aboard the ferry-raft—tapped the destroyer's side and listened intently, with ears pressed against the greasy steel, for answering signals. But not a single tap was heard from inside the vessel. On this score, Captain Calhoun recalled:

"As we shoved off and cut the tail line, Lieutenant Herzinger inquired, 'Captain, shouldn't we say a prayer for those who are not going ashore?'

"So, we stopped for a few minutes and said a prayer—I asked the Lord to watch over the souls of our lost shipmates, to comfort their loved ones and also gave thanks for the deliverance of the rest of the ship's company. By that time, we knew that we had about 30 men missing. We thought that many of them had been trapped within the ship, and that the others had been swept to sea to be either rescued or lost. Divers, later that week, entered the compartments but found no bodies. Accurate figures on our casualties were not to be obtained until early the next day."

Getting the men of the Young aboard the abandoned Chauncey was by far the most arduous and hazardous task. But ferrying them across the flash-floods that ran between the Chauncey and Bridge Rock ran it a close second. And, lastly, the most back-breaking part of the journey: getting the exhausted destroyermen of the Young up the steep, lacerating walls of the rugged rock to its no less rugged top. It was a task in which all men of the Delphy and the Chauncey who still had an ounce of energy left, lent a hearty hand. The Pharmacist's Mates of the two just-named destroyers strove mightily to aid the newcomers.

Alas, Chief Pharmacist's Mate Ralph K. Buchanan of the Young was among the missing, as was Cabin Steward Enrique Torres—who made such ironclad coffee. Also absent were Scotty, Radioman Reddock, Seaman Martin, and Coxswain Salzer.

Among those who proved themselves indefatigable Good Samaritans in assisting injured and/or exhausted men up the cliff was Seaman First Class Patrick Connell of the Chauncey. Of him, Captain Booth wrote:

"When sent to rest and recover from chill, he volunteered to carry messages afoot to Arguello Radio Station, 2 miles distant, over sandy and difficult trails. On returning, he took up rescue work again. Also carried two men, larger than himself, bodily over the cliffs and across a narrow ledge (from Bridge Rock to the mainland) of insecure footing at great personal risk to himself."

A Beacon Lights Their Way

But let us return to Honda mesa, where we left Marshal Bland, Editor Adam, and Policeman Baker after they had sent Captain Robert Morris, ComDesDiv 33, with their SP trackcar crewmen as an escort to the Section House that was the home of Foreman John Giorvas. The trio was attracted to the south side of Point Pedernales as the shouts and screams of a large number of distressed men reached their ears from the direction of Bridge Rock. While they hurried in that direction over the uneven surface of the mesa, they caught up with the small relief party headed by Captain Toaz.

As Captain Toaz told of the grounding of the Delphy, and expressed the belief that perhaps another destroyer was stranded in the immediate vicinity, Marshal Bland (who knew the lay of the land well from past strandings) described Bridge Rock as a barren lava outcropping that was separated from the mainland by a chasm that was fully 20 feet across and connected by a dangerously narrow natural bridge.

Following a brief conference, "Toze" Toaz agreed to wait while Bland and Baker crawled down the slippery cliff to explore.

They crossed the "bridge" and soon found themselves moving among small clusters of men who huddled in groups in the lee of such sheltering rocks as they could find. Using their pocket flashlights, the two lawmen picked their way until they came upon an officer who took them to a spot where Captain Watson and other officers from the Delphy and the Chauncey were helping in the landing of men from the Young.

At that time, Captain Watson's orders, for the men to remain on the rock for possible salvage operations the next morning, were still in effect. The arrival of Marshal Bland and Officer Baker was more than timely because of the great anxiety on the part of the shipwrecked destroyer men to know where the squadron had run aground. Now they received the much-wanted information.

Meanwhile things had been happening atop the cliffs of Honda where Editor Adam remained waiting with Toaz and his party. John Giorvas arrived with a score of men—track workers and bluejackets from the Lee—with several well-tarred railroad ties and a half-gallon can of kerosene. Soon a roaring bonfire was blazing near the edge of the cliff. It gave most welcome light in the grueling operation of getting the injured and exhausted personnel ashore from Bridge Rock. At first, the burden of getting their less capable shipmates up the cliff fell upon the shoulders of destroyer men from the three stranded vessels. But, reinforcements came in the guise of the SP steam shovel and fence gangs headed by Foremen Tobin and Motis from San Luis Obispo.

Equipped with coils of heavy rope, pulleys, stout iron spikes, canvas and wire stretchers, they superimposed their professional rescue skills upon the up-to-then rather haphazard operation. Men who had not suffered too greatly from their experiences were helped up by ropes slung around their shoulders. The slightly injured were carried up the cliff on husky backs. Those who had been seriously hurt were strapped on stretchers and hoisted carefully to the comforts of the life-restoring bonfire. A little earlier, General Foreman Maes had come down from Surf with Dr. M. S. Kelliher and Dr. L. E. Heiges.

They had left their Lompoc offices with such medical materials as they could lay hands on for first-aid treatment. After attending the Lee's men at the Section House for lacerations, the two doctors were taken to Point Pedernales, where many of the casualties were men from the firerooms who had been gassed by smoke following the crash. One enlisted man had a broken arm; another had suffered a broken collarbone.

There were lacerations and bruises by the score. Seaman Eckenberg, of the Delphy, who made the dangerous swim with a life line from the rock to the bluff, had been so manhandled by the seas and the reefs that his body was covered from head to foot with cuts and bruises. Gradually, the men were tranferred from the windblown cliff at Honda and moved through the fog, which now had clamped down tighter than ever, to the section buildings along the railroad track. In describing this scene, editor Ronald Adam wrote:

Hundreds of the men were huddled around the bonfire that had been built there. This was a pitiful sight indeed. Half of the men were without shoes, and some were clad only in their underclothing. Nearly all of them had been drenched in the sea. A cold wind was blowing from off the ocean and the heat from the bonfire was most welcome to the men. Withal, they were the most cheerful lot of men one could hope to find anywhere. They had saved practically nothing of their personal belongings. But as they stood around the fire, they laughed and joshed with one another much the same as a crowd of young fellows will carry on after a football game...

The Mexicans of the section gang had already begun to hold "open house" for the sailors. Every one of the huts, where the Mexican laborers live, had a fire going and cans of coffee simmering. All were full of sailors who sat on bunks, stools, or on the floor and sipped the red-hot coffee...

In the kitchen at the Section House there were as many sailors as could possibly crowd inside. A roaring fire was going in the range and two big cans of coffee were boiling. A sailor who had lost all but a thin suit of underwear was dispensing the coffee as fast as he could but as there were only three cups on the premises and the beverage was steaming hot the service was rather slow.

The kitchen was stifling from the heat of the range and the steam from the garments of the sailors. Several of the boys who had been hurt were curled up on the floor and the others who crowded within the room had been driven there by the cold without.

The living room was almost as crowded as the kitchen. Doctors Kelliher and Heiges were working over the injured. The men who were in a bad way were lying along the walls. They were the first to receive attention from the doctors.

Others with minor injuries were awaiting their turns. The two telephones—one of the railroad company and one of the Santa Barbara Telephone Company—were in this room that was serving the purpose of a receiving hospital. Several of the officers from the ships were here sending out messages when we arrived. Messages were sent out over the Southern Pacific wires and over the telephone company's lines by way of Lompoc. Other messages were dispatched by railroad motor cars to the Arlight Radio Station three miles south. Communication with the outside world was soon expanded through the arrival at Honda of SP linemen and telegraph operators who carried telegraph material and opened telegraphic contacts through new wires they cut into

Giorvas' living room and his two upstairs bedrooms. On reporting to Captain Watson, General Foreman Maes asked what he could do.

"What's needed more than anything else at the moment," said Watson a little hesitantly, "is some food and something hot to drink. But being where we are, and the time of night—[it was now well past 11 o'clock]—I know that it's a fairly large order to fill for some 400 men."

"No," replied Maes genially. "I have been in touch with Trainmaster Foley up at San Luis Obispo. All he'll need is a little time. All your boys will need is a little patience."

"Patience is all we have," replied Captain Watson with a smile that found no reflection in his deeply shadowed eyes. "How long will it take?"

"Oh," drawled Maes, as he looked at his watch, "they won't be able to make the freight train that comes through here about 2 A.M. But there will be time to get the stuff on the Lark, an express that comes through here at four in the morning. Of course, we could send the food and extra help down on a special train!"

"No, no," protested Captain Watson. "There's no need for that. Everything is well in hand. The two Lompoc doctors are doing quick and competent work. I have been thinking; how can we get the men who need hospitalization, there's about a dozen of them, to a hospital?"

"If the doctors agree," replied Maes, "we can put them on the express at 4 A.M. and take them to Santa Barbara. They are well fixed with hospital facilities there."

This plan was approved by Dr. Heiges and Dr. Kelliher. At the same time, Captain Watson put a project into motion that would provide a special train from San Francisco to carry all remaining survivors from Honda southward to San Diego at the earliest possible hour on Sunday, September 9.

Up in San Luis Obispo, action toward participating in the assistance effort at Honda had been building up at a rapid pace. Despite the fact that it was Saturday, and a late hour at that, the town was soon astir with preparations that ranged from sending a special train to Honda to return with a hundred men for hospitalization to dispatching doctors, nurses, medical supplies and food to the scene of the disaster.

As the San Luis Obispo Telegram of September 11 tells the story:

Shortly after the seven destroyers went on the rocks at Honda Saturday night, a call was sent to San Luis Obispo asking that all available physicians and nurses be rushed to the scene of the disaster by a special train, which was ordered to be made up in the local yards. The call also requested that medical supplies, stretchers, and food be sent on the relief train for the aid of the shipwrecked men.

The wire stated a hundred injured men were to be sent here and requested that accommodations for that number be made at the local hospitals. Every physician and nurse who could be reached in this city was notified and all made hurried arrangements to leave immediately. The train was scheduled to leave this city at 3:30 o'clock in the morning, but a later wire cancelled the special but asked that the supplies be sent.

Mayor L. F. Sinsheimer was notified and he immediately began gathering the needed supplies and in a short time taxicabs were rushing the supplies to the Southern Pacific station to be sent south on passenger train No. 76. The groceries were obtained from A. Sauer & Co. and practically all of the restaurants in the city were asked to make hundreds of sandwiches.

The injured men were to be sent here at 4 o'clock but another telegram to the train officials announced that twelve of the most serious cases would be sent to Santa Barbara, while those suffering from exposure were cared for on the spot by physicians and nurses obtained from Lompoc.

Simultaneously, arrangements had gone forward, through Trainmaster Foley's office, to have the generous contributions of the town put aboard the baggage car of the Lark, and he went along himself to take charge of the transportation aspects of the situation at Honda. In and around the Section House, as well as the quarters occupied by the section gang, the population of limping, shivering sailors was steadily increasing as, in groups, they made their way up the sloping mesa and huddled around the four huge bonfires that now blazed along the right-of-way of the rails. These had been built by order of Captain Morris to divide the men into their respective ship's companies. Fortunately, there were plenty of new and used ties piled near the Section House, willing hands to carry them, and kerosense to get them burning.

John Giorvas' little Section House was literally crowded to the rafters. Injured men lined the walls of the living room. The two doctors worked incessantly, their backs bent over John's kitchen and parlor tables, as fresh arrivals replaced men who had received medical attention. The room was noisy, hot, and heavy with the fetid smell of closely packed humanity.

"Wonder what time it is getting to be?" asked Dr. Kelliher of Dr. Heiges.

Just then, a Swiss chalet cuckoo clock on the living-room wall, amid a great whirring of wheels, opened a tiny door. A small bird popped out and repeated its mournful notes a dozen times. It was midnight on Honda and the Lord's day had begun.

7 - Crunched in the Devil's Jaw

The Woodbury's Men Reach Rock

WOODBURY ISLAND, AS IT WAS CALLED AT FIRST, lies about 300 yards to westward of Bridge Rock, which, as related before, is the extreme tip of Point Pedernales, connected to it by a narrow, dangerous natural bridge. In reality, Woodbury Rock could not properly be called an island. It is the top of a small submerged peak, very precipitous and jagged and about 25 feet high. It is not large enough to be noted in the Coast Pilot—as is Richardson's Rock off San Miguel Island—and is surrounded by several detached smaller rocks and reefs.

The range of the tide is only 3½ feet, so that, with great Pacific rollers crashing in, most of its snag-toothed pinnacles are bared whether the tide is high or low. No matter what the tidal condition, great breakers burst into spray against its sides with a monotonous rhythm that sounds like a baleful tocsin of disaster. Captain Davis activated his reserve plan. Namely, to get his ship's company off the disabled destroyer and onto the rock by way of her bow, before her continued pounding on the rocks tore her fragile hull to shreds and drowned all hands. In trying to back out, the Woodbury had widened the distance between her forefoot and the rock where Chief Boatswain's Mate Paul Pointer and four other men, Matthew H. Ryan, Silas A. Puddy, Ralph C. Kiplinger, and Robert M. Wolf, had been standing by since they made their daredevil leaps across the gap between the ship and the petrified lava.

Two lines thrown to them were quickly made fast. Robert G. Warf, Torpedoman 3rd class, was first man over the first line. But, owing to the cutting edge of the rock to which it was attached, the line was severed almost at once. The constant rolling and surging of the ship contributed to this nearly serious mishap. After what seemed an interminable wait to the 90-odd men who stood by, hanging on to any sort of a support, an 8-inch line and three 5-inch lines were put over and made fast, and the evacuation began. By careful handling, the hawsers were secured so as to withstand the ceaseless effort of the mighty muscles of sea and surf to snap them like thread while they heaved and hauled and pounded the defenseless Woodbury.

As Chief Radioman Grover M. Dickman described it:

"The destroyer was settling in the stern. As the thundering breakers struck her, the bow would rise and fall. Then the lines would become taut and strained until the walls of roaring water rolled past. Then they would become slack and sag. As officers and men climbed, monkey fashion, over the line from ship to shore, only superhuman effort kept them hanging on to the snapping hawsers.

"The ship kept settling to port and the deck became impassable. Men could stand on the starboard side of the ship as easily as on the deck itself; they clung desperately to the life lines. Green seas rolled over the stern. To say there was no excitement or that no fear was shown would be exaggeration; however, all hands were eventually transferred to the rock, some a bit short of clothing because the wreck caught them in their bunks. There was much rejoicing when it was reported that all hands were safe in spite of the fact no one knew where they were."

While the abandoning of the ship was in progress, N. F. Owen, Quartermaster 1st class, and H. F. Williams, Seaman 2nd class, who were standing to on the forecastle, thought they heard a cry of distress from the sea off the port bow. They listened. Heard the call again.

At the same instant, a strong torchlight from farther aft picked up a man who was making feeble swimming motions in the water. Tying one of the discarded life lines around his waist, Williams sprang into the sea on a passing surge of surf. He reached his man and, with the aid of Owen, pulled him aboard the Woodbury. Resuscitation methods were promptly applied and, after his lungs had been cleared of sea water, the rescued swimmer was found to be Boatswain's Mate Second Class Braden, who had been washed overboard from the side of the Young. Commander Davis and Carlos, the Wardroom's Philippine Mess Attendant, were the last to leave the ship. Carlos had been confronted with a problem soon after he, with all his shipmates, had rushed on deck. In his hurry to get topside, he had forgotten his savings—all in cash and said to amount to $2600. As the evacuation neared completion, Carlos sneaked below while none of the officers was looking. He did this despite the warnings of shipmates who were desperately intent upon saving their lives and not worrying about what had been left behind. They were grateful just to be alive.

Nevertheless, Carlos disappeared and did not return until all but Captain Davis had left the ship. Then he went across on the hawser, hand over hand like the rest. As the lad swung toward the rock, it was noticed that his trousers had been tied at the ankles and that they were bulging as if filled to the bursting point. "Hey, fellows," piped one of his shipmates, "look at Carlos—old Mr. Moneybags himself. Them pants! They're busting with greenbacks!"

But Carlos just grinned like an ancient idol. Instead of looking for mere money, he had grabbed something of greater import than cash to castaways of the sea. His pants were chock full of oranges. Another man who wanted desperately to get below on a salvage job, but realized the hopelessness of his mission, was Ensign Paul Howell. Recently married, he had undertaken to transport all the more bulky wedding presents, silver and so forth, to San Diego. His bride never saw them again—that Ole Debbil Sea—or looters—got the lot. The Medical Officer for DesDiv 33, Lieutenant H. W. Miller, who was quartered aboard the Woodbury, was one of the last to leave the ship. He carried with him a first-aid bag for which, fortunately, he had little use. Most men had managed to get their shoes and clothing on, so cuts and abrasions were at a minimum. The good doctor had been stretched out on the wardroom transom, absorbed in Victor Hugo's Toilers of the Sea.

He had just reached the point where the hero battles with a giant octopus when the ship struck and loose gear came tumbling over him. His messmates say he came to his feet, white of face and yelling. He thought the octopus had him! But disembarking from the Woodbury was a well-organized retreat and not a rout. Large quantities of bread, cooked meat, and canned goods—as well as water—were transferred to Woodbury Rock. As the warmth of a bonfire would also be needed, Captain Davis, with excellent forethought, ordered the vessel's gangplank pulled across and broken up with axes for firewood. This task took a lot of doing, but—since every man sweated in everybody's interest, including his own—it was finally accomplished. Old paint buckets and tar pots, salvaged from the paint locker in the bow, further added to the welcome light and heat.

It was some minutes before the heavy wooden gangplank could be broken up and turned into bonfire fodder. This was needed as a starter for the tar pots and old paint buckets. Until that welcome moment, the men huddled together to keep warm. They were also warned against moving around in the dark, lest they should stumble into holes or plunge into deep crevices. The fog was so thick that it had become a moisture-soaked blanket, but it was no colder than the chill sweat of anxiety that ran from the pores of many of the shipwrecked men. But here, as elsewhere, courage and training plus discipline held emotional pressures under firm control.

Near Death in a Whaleboat

At the time she ran aground, the deck force of the Fuller rigged a collision mat over the gaping holes in her side, while fireroom and engine room crews remained at their stations and pumped desperately until they were actually driven out by the rising water. By the time boilers were secured the men virtually had to swim for the ladders. They were: Valerius Vasvinder, Watertender 1st class; Frank C. Colpitts, George L. Lord, James Moore, and Harold Davis, all Firemen 1st class. The collision mat was no more successful in stopping the inrush of the sea than Canute's command was in halting the tide.

Soon Captain Seed decided that, since the prospects looked hopeless, steps must be taken to ensure the eventual evacuation of the ship's company.

He ordered Lieutenant (jg) Homer B. Davis, his Executive Officer, to lower the whaleboat, which was fairly well sheltered on the starboard side of the well deck, and man it with volunteers. Ensign Bascom S. Jones was placed in charge of the boat.

Captain Seed's plan was to transfer the Fuller's men to the Woodbury, which obviously was in a much better position than his own destroyer.

The Fuller was pounding badly, and there was also the ever-present possibility that she might slide off her supporting ledge into deeper water or even break in two. The mission of the whaleboat was to run a hawser to the Woodbury and thus set up a ferry between the two ships. To do this, it would be necessary for the boat to make its way out of the encircling rocks, reach the Fuller's port side, take the end of a line, row to the stern of the other destroyer, secure the hawser, and return to the Fuller. It sounds like an easy mission, especially if one repeats it rapidly. While the crew was being selected from the volunteers and the ship's lifeboat made ready, Ensign Jones made a quick tour of inspection around the deck and found that the Fuller was wedged into a nest of rocks with a semicircular reef reaching from her starboard bow to her stern. On the port side, a large rock showed amidships just below the surface. The razor edge of the port propeller cleft the water at the stern. A heavy ground swell, coming in over the stern, made tremendous odds against launching a small boat successfully on the port side. And getting through the semicircle of rocks to starboard appeared to be nothing short of suicidal.

After a brief conference with the Exec. and Captain Seed, Ensign Jones, a bronzed young Georgian with the keen features and sharp eyes of an athlete, took his place in the whaleboat where his volunteer crew already were on their thwarts. The men selected were Frank Stasta, Casey M. Bass, Lloyd A. Campbell, and Alfred M. Titus, all Seamen 2nd class; George H. Trenhaile, Fireman 1st class; Sumpter L. Gillespie, Fireman 3rd class; James R. Collins, and Glenn M. Melvin, respectively Fireman 2nd class and Shipfitter 3rd class. But let Bascom Jones spin his own yarn about that perilous odyssey. He was a man endeared to his shipmates by his readiness to lend a hand wherever needed and by the determination which, in spite of many difficulties, had won him a diploma from the Naval Academy. If sheer grit could do a job, Jones had it made. His story begins as those on deck begin to lower the whaleboat down the starboard side of the rolling, pounding Fuller. For the next few minutes everything seemed to go contrary to all the laws governing the handling of small boats. While still suspended several feet above the water, the block at the stern was tripped by an especially vicious wave and that end of the boat dropped into the sea.

Theoretically, the waves rushing in from the stern should have swamped us immediately, but quick thinking by the man on deck saved us from a dangerous ducking as he immediately cast loose the forward fall.

Still, this little mishap was serious enough, for in the drop, our only light was smashed and the "crutch" or oar lock for the steering oar was damaged beyond immediate repair.

At this one time, when accurate steering was needed to get through dangerous rocks, we were left to the mercy of the sea. Afraid of catching under the guard rail of the destroyer and being swamped as our small boat rose and fell with the waves, I gave the order to "shove off in the bow."

Here again our luck seemed to defy the incoming waves, and as our bow swung out we were able to drop our oars into water and pull away from the side of the ship.

And now in darkness, unrelieved by lights of any kind, and a running sea we were faced with the problem of slipping through the jagged teeth of the reef, without the help of a steering oar.

I had picked, as our point of exit, the only passage I could see in the whole line of rocks, a passage in the center of the reef between two rocks, which offered a slight clearance on either side of the boat. Trusting to Providence and the sharp eyes of two men in the bow, we headed for the passage in a remarkably zig-zag manner. The unevenness of our course was due to the steering, accomplished by having one side row hard to turn the boat in one direction with the side reversed to turn in the other direction.

If we were lucky we might hit the passage, otherwise there would be just one less small boat in the navy. Fortunately, the oars were handled by men trained to obey orders quickly and confidently. Making the best speed possible we headed for the rock on the seaward side of our passage. A few feet from the rock we tossed oars, held our breath and waited to see if our guess was right.

A wave, like the one which had come so close to being our undoing, now seemed to try and make amends for former unpleasantness, for, with a rush, it lifted us up, moved us over, and dropped us in the center of the passage through which we shot without even scratching the paint.

After trying unsuccessfully to get alongside the ship on the port side we were ordered to cease trying, for a touch of the propeller or a scrape against the submerged rock would have sent us to the bottom, and the sea was running in such a direction as to make it impossible to come alongside without a steering oar.

Our only problem now, was to keep afloat in the vicinity of the ship until daylight, in the hopes of picking up a few men in case the ship went down.

We pulled out to sea to get clear of the rocks, and drifted with the current. We could not drift long, as the night was cold, the men insufficiently clad, mostly in underclothes and life jackets, and rowing was necessary for warmth.

But which way to row? The few flashlights, at first visible on the stranded destroyers, had disappeared as we pulled away from the rocks, and, in that darkness, we could have rowed to sea without knowing it. I knew the direction of our drift relative to the shore but which way was shore?

Anxious to keep close to the ship where our help might be needed any minute and worried about the men, who would not share my warmer coat and who were getting colder all the time, I finally gave orders to out oars and give way together.

Fortunately, at that moment, from out of the darkness, came the faraway wail of a train whistle. It was a weird sound in the night but a most welcome one. Plotting his relative position by the sound, Jones was soon back at his ship although no one sighted the whaleboat.

Later on a fire was lighted on the big rock, by the men from the Woodbury and this light acted as a beacon to guide Jones throughout the rest of the night.

8 – Midnight on the Mesa

Mixed News from Point Arguello

SOON AFTER MIDNIGHT ON THE MORNING OF SUNDAY, September 9, the destroyermen on Honda mesa were organized into ship's companies and rough tallies taken of those present or accounted for. The Captains of the four vessels, the Delphy, S. P. Lee, Young, and Chauncey, whose men had clambered up the cliffs from the sea, like lemmings reversing their migration, reported the results of their roll calls to Commodore Watson and Captain Morris in the headquarters established by the former in the unused upstairs bedroom in the Section House.

Lieutenant Commander Hunter of the Delphy, reported two men known to be lost or missing, with one still aboard the ship. Commander Calhoun of the Young had about 30 of his crew missing and unaccounted for. As he submitted his tragic summary, Bill Calhoun felt a hot, prickly sensation behind his eyes and there was a bit of a tremor in his voice. But he maintained his disciplined bearing and looked steadily at the Squadron Commodore whose usually ruddy face had lost its glow and whose own eyes were sick with misery. A planning conference followed the submission of the roll calls. While it was in progress, Robert Sudden, better known by the old Spanish title of "Don" Sudden, entered the room. It will be recalled that his home, Bancho La Espada, stands but a few miles south of Point Arguello Lighthouse and that Honda mesa forms the northern end of his property.

Mr. Sudden explained that his wife Lita had been awakened earlier during the night by the constant ringing of the telephone party line. Curious to know what could cause such unusual activity, she had slid out of bed, gone into the living room, lifted the receiver off its hook and learned that there had been a big shipwreck up at Honda. Aware that fog had descended upon the local coastal region and, knowing that Arguello Light did not have a telephone, Mrs. Sudden awoke her husband. She suggested that he should drive up to the Lighthouse to make sure that Keeper George Olsen had this important information.

It took but a moment for Mr. Sudden to throw on some clothes, get into his car and drive over the rough and bumpy Sudden grade that led to the Light. Point Arguello Lighthouse stands on a rocky bluff that towers steeply some 100 feet above the restless surf. Approaching slowly through the fog, Sudden was amazed to find the windows in Keeper Olsen's house aglow with light. On entering, he found the Keeper, Mrs. Olsen, and First Assistant Keeper Arthur A. Settles, plus a couple of men from the nearby Navy Radio Compass Station, administering first aid to five bedraggled and bruised sailors whom they had just hauled out of the sea.

The men were in bad shape and in need of immediate medical attention. Luckily, Mr. Sudden recalled that his wife—in recounting what she had heard over the telephone—had mentioned something about doctors being at work on survivors at Honda.

The men were much too dazed and sick with shock to stand the rigors of an automobile ride across the mesa, so Robert Sudden offered to drive up to the Section House for medical assistance.

"Five men!" ejaculated Captain Calhoun as he heard the news. "They must be mine. Did any of them say what ships they were from?"

"Yes," replied Mr. Sudden. "One of them, he said his name was Scotty, told me that they were all from a destroyer called the Young."

"Thank God for that, thank God for that," exclaimed Calhoun earnestly. "Here, I'll take you downstairs, Mr. Sudden, and get one of the doctors to go back with you to Point Arguello Light."

The story of the rescue of Scotty and his four companions, as reported by First Assistant Keeper Arthur A. Settles is a precise account that does not deviate to underscore or give color to an incident of rather heroic proportions. It follows:

The (lighthouse) crew was aroused about 10:30 P.M. by cries for help. In the thick weather only a feeble light was discernible to seaward. Whatever it was, no landings could be made on the very steep and rocky north shore of the point because of the high surf. Whatever it was, it would be sure to be dashed to splinters on the rocks by the huge breakers. With shouts and lanterns, we tried to make them understand that they should try to lay to on the southern side of the point where the shore is protected and the surf less violent. It worked. The light slowly—it seemed hours—fought incoming surf and outgoing swells and finally moved around as designated.

Presently, the light vanished but in a little while we heard calls and yells below the point. Meanwhile we had sent for helping hands from the Naval Radio Compass Station nearby. Ropes were lowered along the virtually perpendicular cliffs—about 75 feet high at the lowest point—and a couple of sturdy young sailors were sent down when we found that the men were not able to tie themselves to the ropes. We hauled them up one by one. The poor lads were in terrible shape. Numb with cold, drenched with crude oil and completely exhausted. Had their lives depended on it, the five of them could not have taken a single step or crawled the distance of a yard.

They were utterly spent. After we got them into Mr. Olsen's house and took such care of them as we could, until they could get medical help, we learned that there had been two rafts held together by the grab ropes that went around their sides. The other raft was not on hand. There were three men in it.

Then one of the men told this story:

The two rafts had but one paddle between them. When it was found that the one paddle was not adequate to fight the two rafts in the eddies that swirled around the point, the three men aboard the four-man raft had let their hand-hold go.

One man shouted: "Make a good landing, Scotty!" and that was the last the men on the rescued raft saw of them. It was a deliberate sacrifice on the part of these three men that their comrades might survive.

Nothing was ever heard of this raft. Chances are that it drifted south—was hurled upon the rocks of one of the Channel Islands and every man on board was pulled to sea by the dreadful undertow. Fireman First Class F. Scott, believed lost and credited with great courage in standing by to close the master valves that fed the Young's burners, was one of the destroyermen on the lucky raft.

The others were James Stallman, who was later hospitalized, and Schick, Kolb, and Zobria. The three men whe were lost on the raft that drifted out of sight were Charles A. Salzer,

Coxswain; Clitus A. Reddock, Radioman 1st class; and James T. Martin, Seaman 1st class.

In telling his story, Fireman Scott explained that, through God-given strength, he had been able to open the airlock scuttle when the Young's fireroom was being flooded by inrushing sea water. He was literally sucked into the sea as he reached the slanting and submerged deck. He had grabbed some piece of flotsam and for a while—he did not know for how long—he was out cold. When he recovered consciousness, Scott was in smooth waters and deep silence.

After what seemed an hour, Scotty bumped against something smooth and soft. It was a raft with Stallman, Kolb, Zobria, and Schick aboard. They hauled him in. There was no visibility. And, for a while, the only noise was the sound of distant surf that seemed to come from the left. Just on general principles, all five began to shout. Listen. And shout again. Soon they were rewarded by a reply. What with only one paddle aboard the raft, they began to move awkwardly in the direction of the voices. Soon they came upon another set of drifters. Three men in a raft without a paddle to their names—Reddock, Salzer, and Martin. One of them took hold of the grab ropes on Scott's raft. And now eight men sat silently on two rafts on the surface of the ocean, pushed by seas driven by the northwest wind. Gradually, the sound of the surf grew greater. Soon they could feel the pull of the outgoing ground swells. Knowing that they must be near land, Scotty, the most durable member of the quintet, began to paddle toward shore.

Miraculously, at this very moment, the fog lifted. Their eyes caught the beam of Point Arguello Light. Their ears heard the intermittent deep-throated growls of its foghorn. Working the paddle with frantic fury, Scotty headed toward the Lighthouse atop its tall cliff.

As Scott paddled, Stallman flashed his pocket light. On a preceding page, First Assistant Keeper Settles told of the rescue of Scott and his companions, as well as how Reddock, Martin, and Salzer played long odds against their own survival to improve the chances of their shipmates. The courageous and self-sacrificing gamble taken by these three unsung heroes did not pay off. Air sweeps conducted during the following days by Navy pilots failed to disclose any trace of the raft. On Tuesday, September

11, a ship reported sighting something that might be a raft 7 miles north of San Miguel Island. Despite an all-ships radio sent out by Admiral Kittelle, the raft and its luckless passengers were never spotted again. Most seagoing men were of the opinion that the raft was carried to destruction on one of San Miguel's outlying reefs and its three occupants sent to their deaths. Soon after that midnight roll call was completed along the rails on Honda mesa, Captain Watson prepared a duplicate dispatch to the Commandants of the Eleventh Naval District in San Diego and the Twelfth in San Francisco covering the extent of the disaster as known to him at the time. Since the messages were too important to be read over the telephone to Charles L. Atkins, the Third Trick Operator in the SP dispatcher's office in Surf, Lieutenant Lawrence Wild, Squadron Communications Officer, hiked down to Surf by walking the sleepers on the railroad track. At that time, Captain Watson could not, with any certainty, report firmly on the stranding of more than four destroyers and the status of their complements. Communications were greatly improved during the night when SP linemen completed the installation of additional telegraph lines. Soon railroad, as well as Western Union, telegraphers arrived to handle the steadily mounting traffic of incoming and outgoing official messages.

A special communications problem arose from the quickly swelling numbers of officers and men who wanted to reassure their families that they were safe. Others wanted to use the Santa Barbara Telephone Company's party line, which was being kept clear for official use by City Marshal Bland. The lawman had asked for and received the cooperation of local subscribers in keeping the line open for calls to and from Honda on Navy business—and there was a lot of it. Among the many San Diego men who were eager to reach their homes by phone was Chief Electrician's Mate Kerrigan of the Young. Like any other expectant father, under the circumstances, he was on pins and needles with frantic anxiety. He had to let his wife know that he was safe but that he would not be home on Sunday as planned. Marshal Bland, kindly but firmly, told the burly Chief that the wire was for official calls only. In his despair, Kerrigan turned to Captain Calhoun for assistance.

He gave it without stint. Chief Kerrigan received Captain Watson's approval to place a call. Even so, it was not until 7 A.M.

that the frantic father-to-be had the calm little mother-to-be on the telephone.

"Well, what did she say?" asked "Red" Hall who, in a manner of speaking, had been holding Kerrigan's hot and hairy hands during the long hours of waiting.

"Oh," replied Kerrigan, "she says she can wait until tomorrow. There's no hurry. No hurry? And me chewing my fingers off!"

"Gawan," spoofed "Red," "you're only sore because it won't be born on Admission Day."

"That baby isn't any 'it,'" retorted Kerrigan firmly. "Admission Day be hanged. There's an eclipse of the sun at noon tomorrow. And, believe you me, my kid is gonna eclipse the eclipse."

And it did.

To dispose of the message situation at Honda, and its nationwide effect, it is estimated that more than 700 personal telegrams were written and delivered for transmission to almost all 48 states before the destroyer men entrained for San Diego that Sunday afternoon. As they piled up, scrawled on scraps of paper, the messages were forwarded to Surf, Lompoc, San Luis Obispo, Santa Maria, and Santa Barbara for transmission over Western Union and Postal Telegraph wires.

A great majority of the telegraphic revelations of narrow escapes from sudden death and great disaster reached the families of these men before they had heard a word about the Tragedy at Honda. Therefore, the telegrams set off chain reactions of emotional outbursts that reverberated through the populations of hundreds of communities and, eventually, swept the entire country.

Where was Honda? What had happened at Honda? What was the truth and what was being hidden from press and public about Honda?

Truthfully, nothing was being hidden. It was simply a case of the facts in the matter being slow in their unfolding. Again, it was the unfortunate timing of the disaster that was responsible. If the jinns of Honda had conspired to set up a veritable tidal wave of confused thinking, they could not have wrought better than they did by the strandings taking place on a desolate coast at a late hour on a Saturday night. With characteristic promptness, the press and public wanted action, once the news of the disaster became known. But Honda, being what and where it is, was not readily accessible for rapid action of any kind except seaborne disaster.

Angel of Mercy on Fast Freight

When the SF's southbound fast freight reached the Section House on Honda mesa about 2 o'clock in the morning, a plump, motherly-looking little woman dropped off the caboose and smilingly made her way through the crowds of sailors who milled around the bonfires near the track.

In a large market basket, she carried several dozen eggs, quarts of milk, pounds of coffee, and loaves of bread. She was Mrs. Charles L. Atkins, wife of the Third Trick Operator at Surf.

Many willing hands were prompt to carry her promising-looking burden for her. After hearing about the disaster, she had set about getting a neighbor to mind her two young children, collected whatever food could be spared from nearby kitchens and arranged with her husband to go up to the mesa on the 2 o'clock freight.

As it happened, Mrs. Atkins was the first and only woman who came to Honda that Sunday morning for the purpose of making herself useful.

But whatever plans Mrs. Atkins—the boys were soon calling her "Ma" Atkins—may have had in cooking chow for some 400 hungry sailors out of even a large market basket, were suddenly abandoned.

The moment the two Lompoc doctors laid eyes upon her, as she entered the temporary hospital set up in Giorvas' little parlor, "Ma" Atkins was drafted to look after the injured men. Despite the competent assistance given by the Pharmacist's Mates of the S. P. Lee, the Delphy, and the Chauncey, there just were not enough hands to perform all the services for the injured that needed doing. And, at a moment like that, there is no known substitute for a motherly woman.

So, instead of taking command of the kitchen in the Section House, Mrs. Atkins turned her attention to assisting in the care of the injured. Of these, 13 men from the Delphy and 5 from the Young required hospitalization. It was arranged that when the southbound Lark arrived at Honda about 4 o'clock in the morning 10 destroyermen of the Delphy and 3 of the Young would be taken on for hospitalization in Santa Barbara. The remaining were to be taken to the Naval Hospital at San Diego on the special train which was to arrive later in the day.

The caboose of the fast freight had held other blessings for the men at Honda than the arrival of "Ma" Atkins, her gentle smile, tender hands, and market basket. Cold and wet sailors crowded around as brakemen unloaded great heaps that contained blankets by the score and comforters by the dozen—even overalls, trousers, coats, and sweaters.

These welcome gifts, immediately placed in the vigilant custody of Chief Petty Officers, had been hastily donated and sent by residents of Lompoc. In fact, most of the blankets and comforters had been contributed by the Moore Mercantile Company of that town.

To be sure, there was not enough to go around. But the neediest cases were taken care of. And soon, oil-smeared sailors, who had weathered the fog-dampened night hours of Honda in little more than their underwear, sat or strutted around the fires with comforters or blankets draped around their shoulders Indian style. As for footwear, they wrapped their feet in strips of blankets and took a new lease on life.

Never was the shipmate spirit, that is typical of small ships, better demonstrated than on Honda mesa that night.

Officers and enlisted men alike took care of their less fortunate comrades—shared their clothing, shared what was even more precious—smokes. Concern for their shipmates was first in everyone's mind. Greatest of all concern was displayed by Captain Watson. He seemed to be everywhere at once. The weight of complete responsibility was on his back and it drove him with relentless compulsion. Because the responsibility of a commander cannot be shared, it is a cruel and relentless driver. And the Squadron Commander's haggard face gave ample proof that he was in its toils.

The hundreds of men who milled around on Honda mesa in a chill, damp wind formed a picturesque and gypsy-like, yet tragic community. Many of them, despite the energy and resilience of youth, ambled about like old men. Their exhaustion was a burden that bent their shoulders and hobbled the movements of their feet. Most of them continued to take their misfortune as just sailor's luck—which is either good or bad.

Some were morose. Some were merry. All were hungry. But, since the word had been passed that food would arrive at 4 o'clock in the morning, they took the period of waiting with fairly good patience. Now and then, a quickly swelling and quickly fading chorus would sweep the scene with the call: "When do we eat?"

While it sounded much like the roar of a pride of famished lions, it was wholly free from ferocity. Shortly before the Lark arrived about 4 A.M., the men bound for Santa Barbara were made ready and placed in charge of Lieutenant Allen P. Mullinix of the Delphy and Ensign Robert Greenwald of the Chauncey, Communications Officers of their respective ships.

Lieutenant Mullinix, who came out of the wreck almost in the clothes nature had provided, plus a badly injured leg, was able to borrow a pair of dungarees and an overcoat from some of the more fortunate survivors and made the trip to Santa Barbara and on to San Diego in this attire.

On reaching the Destroyer Base, he was hospitalized. Getting the 13 men aboard the train was accomplished in not too many minutes by Foreman Tobin's SP gang. At the same time, members of Foreman Motis' outfit tended to the unloading of the boxes of food and medical supplies that had been collected in San Luis Obispo. Also aboard the Lark was Trainmaster Foley. He reported to Captain Watson that a special train, to be assembled in San Francisco, would reach Honda in mid-afternoon.

The Squadron Commander had hoped that the departure might take place somewhat earlier, but Foley explained that train schedules on the Southern Pacific and the Santa Fe railroads made this impossible.

Ordinarily, Southern Pacific passengers, bound for San Diego, would change trains in Los Angeles. In arranging for a through train over both systems this major inconvenience was avoided.

All hands were duly grateful. Now, with the departure of the most seriously injured aboard the Lark, "Ma" Atkins turned her attention to coffee making and the distribution of the newly arrived food.

In addition to several hundred ready-to-eat sandwiches—restaurants, householders, and food shops in San Luis Obispo had provided sliced and unsliced meats, sausages, cheese, and bread as well as liberal amounts of butter, coffee, salt, sugar, and canned milk. Plus: Cigarettes!

"Ma" Atkins' problems with respect to providing solid food for her hungry charges had been solved for the moment. However, a quick estimate revealed that there would not be enough for second helpings. No one would be overlooked.

On the other hand, no one would eat his fill. Then there was the matter of coffee. Trainmaster Foley had borrowed a couple of half-gallon coffee pots from work-train kitchen-cars in his yards and added a few dozen heavy coffee mugs. That was a great help. But the making of coffee and serving it to some 400 men would be a slow process that demanded a lot of patience by all concerned.

So "Ma," with the aid of commissary stewards, cooks, and mess cooks, set about to make sandwiches and brew coffee. The latter was made over red-hot embers along the edges of the bonfires. By and by, the aromatic odor of newly made "Jamoke," to use Navy slang, spread over the improvised camp along the railroad track.

This simple proof that hot coffee and food was in the offing lifted the chins of many downhearted men. Soon the survivors of the Young took the lead in general singing. They led off with their theme song: "Yes, we have no destroyers, we have no destroyers today!"

As Editor Adam observed: "They sang in true jazz style." On the subject of "chow" he had this to say: "Although the boys had to stand in line and wait, as service was necessarily slow for lack of cups and adequate cooking utensils, the time finally came when they had at least a cup of coffee and a sandwich with what was better still, a cheery word from Mrs. Atkins."

Chances are that a cheery word and a warm smile from motherly Mrs. Atkins to each man, as he advanced with the chow line, was as welcome as the food, coffee, and cigarettes she distributed with quiet efficiency.

CARLEY RAFTS (OR FLOATS) VISIBLE ON HMS RODNEY

9 - Dawn in the Graveyard of Ships

The Fuller Gives Up the Ghost

SHORTLY AFTER THE FULLER'S WHALEBOAT VANISHED into the night without leaving any clue as to its safety or destruction, Lieutenant Homer Davis obtained Captain Seed's permission to try to take a life raft to the Woodbury for the purpose of carrying a line between the two ships.

As previously described, there were about 100 yards of extremely heavy surf between the two destroyers. To launch the raft, it would be necessary to put it over the wave-swept port side and the odds against success were heavy. Still, if the Fuller failed to stand up against the severe punishment she was receiving from the rocks that prodded into her bottom and sides, the possible sacrifice of a few would be overbalanced by the rescue of the greater number.

Still, there was—thought Captain Seed—an outside possibility that a raft might make it. Lieutenant Davis was backed up by a trio of volunteers who would be sturdy seconds in any wind-swept corner. They were Chief Boatswain's Mate H. A. Thomas; Boatswain's Mate First Class Ollis Kelley, and Torpedoman Second Class Walter Scott. With swift but careful consideration, Captain Seed gave his consent. After considerable difficulty, a raft was launched.

As it bounced alongside the vessel, the rubber craft bucked and reared like a wild stallion trying to escape a horse wrangler's lariat. But, despite its unpredictable gyrations, Davis, Thomas, Kelley, and Scott succeeded in leaping aboard.

A line had been bent to the raft's stern so that it could be hauled back, once contact had been established with the Woodbury. Equipped with paddles, the four men buckled down to the herculean task of making a 100-yard journey through heavy surf and receding swells interspersed with crashing breakers. It just could not be done. The raft, reduced to a plaything of the seas, spun and yawed without the slightest response to the paddles.

A heavy comber lifted it and turned it over as deftly as a short-order cook flips a pancake. The four occupants tumbled into the sea. It was only through quick thinking, supreme physical stamina, and the hand of God that each of them fought and found his way to the overturned raft.

They gave the haul-in signal on the attached halyard as they clung breathless and exhausted to the grab lines on the raft. In time, they were once more aboard the Fuller. Aboard the destroyer, all hands had received life preservers and were brought forward of the galley deckhouse. As the hours of the night dragged by, the wind and sea gradually worked the vessel ahead until the bow was jammed between two rocks. At the same time, she settled so heavily by the stern that, just before daylight, the water had crept up the slanting deck to the base of the searchlight structure.

From his bridge, Captain Seed kept anxious and vigilant watch. No doubt about it, the feel of the Fuller, according to the way she was working, was that she might turn over without warning when the now-reversed tide came in stronger.

Among the manifold worries that had weighed heavily on Captain W. Dudley Seed—"Dud" to his classmates—that night, was the possible loss of the men aboard the whaleboat that had vanished into the night.

Having no means of signaling, Ensign Jones had been unable to let the ship know that he and his men were still afloat. Miserable as they were, the destroyermen aboard the Fuller gave three rousing cheers when the whaleboat hove into sight like a gray ghost in the gray dawn.

After ascertaining that the men were in good shape, Seed, shouting through a megaphone to make himself heard over the deafening noise of wind and wave, asked Ensign Jones if he could make a try at taking a line over to the abandoned Woodbury. Wet and shivering though they were, there was still a lot of spunk left in them and Bascom answered with a wind-blown but hearty, "Aye, aye, sir!"

Strengthened by the sense of their new responsibility, the men in the whaleboat heaved on their oars with renewed vigor, reached the Woodbury, and, in time, bent a hawser to her after whaleboat davit so that a breeches buoy might be rigged. But, unluckily, after taking a strain, the bow of the Woodbury started working off with the threat that the vessel might slide astern and into deeper water. The plan had to be abandoned. Next, the heaving line was passed by the whaleboat's crew to the top of Woodbury Rock where the other destroyer men had sought safety.

This done, Ensign Jones and his men clambered up the slippery, canted deck of the Woodbury to the bow and across the now steel-taut life lines, hand over hand, to Woodbury Rock. As bad luck would have it, this heroic straggle on the part of the whaleboat's crew did not pay off either. The hauling line to the hawser carried away just short of the hands of the men on Woodbury Rock. Captain Seed ordered Jones and his men to remain where they were.

Once more, the Fuller was without an avenue of escape. And as time ticked off, the seas surged higher and higher along the stranded destroyer's deck until their white, murderous claws curled about the base of #4 smokestack. With heavy seas breaking over the after part of the ship, an occasional wave cascaded over her forecastle.

Forward of the galley deckhouse, men sought sheltered places where they could on the four-stacker's well deck, unprotected by bulwarks.

"The men lay in groups," recalled Captain Seed in his report. "Two of these groups sang for a long time. The most popular song seemed to be 'Where do we go from here?'"

Captain Seed was standing at the bridge ladder, considering his next course of action, when Frank M. Moon, Machinist's Mate 1st class, approached and said, "Captain, I've got another idea."

"That's fine, Moon. Shoot! What's on your mind?" replied the Captain with an interest that revealed his anxiety. This was not an occasion for polite formalities. Straight from the shoulder is the small-boat approach to problems. The Fuller's captain had not forgotten Moon's courage and initiative in taking a line to a small rock soon after the Fuller stranded.

That the rock was not only too small, but also would be under water with the rising tide, was not Moon's fault.

"Well, sir," began the Machinist's Mate, "there's only seventy-five yards of water between our port bow and the big rock where the Woodbury men are. I'd like to try to take a line across."

"Do you consider yourself a strong swimmer?" asked the Captain. "Yes, sir," replied Moon without hesitation; "I've had my share of surfing and rough water. I can make it."

Dudley Seed's gaze turned from Moon's determined features to the no less ragged face of the shore line, barely visible across a gap of water so rough that it boiled and bubbled like the brew in a witch's caldron.

"Tell you what," grinned Seed as he slapped Moon's shoulder. "I'll see if I can swim across—and if I can, you follow with the line."

Lifting his head, Dudley Seed looked up at Lieutenant Homer Davis, his Executive Officer, who, from an open window on the bridge, had overheard the conversation. Before Davis could break in with a word of remonstrance, Seed said:

"You heard it? Okay! If I make it, send Moon along with a line and have the crew in life jackets, haul themselves over the line to the rock one at a time. If I do not make it, good luck and you're on your own! But a word to the wise: Don't you ever paint that wardroom again!"

With this, Captain Seed set aside the cap on his crisply blond head, removed his foul-weather coat and slipped off his shoes.

As an afterthought, he tossed his cap up to the Exec. "You may have to wear two hats," he said grinning. Then with a running start, he dove into the sea in order to surface as far from the oil-slicked side of the ship as he could. It was a long and rough 75 yards for even a professional channel swimmer. Dudley's years pulling an oar in the Naval Academy's championship crew stood him in good stead. He had a magnificent physique and, although he took a terrific beating from the sea, he made it to the rock.

And so did Moon. Ensign Jones and his men were on hand to take charge of the line and land the crew as they came across from the Fuller. To start with—because of one of those foul-ups which at times seem as unexplainable as they are inevitable—the evacuation did not run too well.

As Ensign Jones explained the situation:

While I manned the rock end of the line over which the crew of the Fuller pulled themselves, the water level was about 2 feet below me. As each man scrambled over the line, I waited for the water to rise, reached down and caught each man by his life jacket and hauled him to safety. Most of the men were exhausted by the time they reached the rock—not only from the long night and from the tension and exposure, but from the long fight of pulling themselves over the line from the ship.

The waves were heavy and the cross currents were very strong. As high above the water as I was, I was, nevertheless, frequently covered by waves. The currents seemed to run in both directions. So strong were they that they slapped the men, who crossed the line at arm's length, in one direction then, changing, slapped them the other way. The simplest way of getting the men across from the Fuller would have been to have my whaleboat crew haul each man across and let the Fuller haul the line back for the next man.

However, I was unsuccessful in getting this done and all but about eight or ten officers and men jumped from the Fuller into the water and pulled themselves across, hand over hand, until—with the help of the incoming wave and the line—we could reach down and catch the back of the life jacket and pull them up the face of the rock to safety.

When the men were in the water, the good strong swimmer would come along quickly hand over hand; but the poor swimmers, and those physically exhausted from the long night of hanging on to the sloping deck of the Fuller, would get halfway and then find the going very tough and would have to be almost mentally pulled the last few yards to safety.

One case in mind was the Wardroom Steward, a very large man. With a line under each arm, it was practically impossible for him to slip off the lines. Still, he froze to such an extent that he could make no move except to yell:

"Save me, Mister Jones, save me, I'm drowning!"

After a considerable amount of cussing and urging, he crept inch by inch over to the rock and was dragged to safety. Almost immediately afterward, Ensign Sweeny jumped into the water just as the waves receded. He hit a rock with enough force to knock the wind out of him. He was dragged back on board just in time. I was then able to get my request granted. By using a bight about the waist of the remaining officers and men we were successful in getting them to the rock quickly and safely. Fresh fuel was added to the bonfire on Woodbury Rock as the early settlers on this bleak haven from the diabolical sea made room for the bedraggled newcomers from the Fuller.

Almost dry, almost comfortable—thanks to the twin blessings of food and fire—they passed hunks of bread, chunks of meat, plus cans of fruit and vegetable juices among their welcome guests.

One of the soaked-to-the-skin, chilled-to-the-marrow, and hungry-as-a-lion Fuller men, said with deep admiration, as he dried out, melted out, and filled out: "Gee, but you Woodbury b—ds really travel first class when you hit the beach!"

Then he slumped on his side, blissfully adrift on the smooth seas of dreamland. He was also completely fagged out. The Jinns Lift the Curtain Slowly Captain Robert Morris, ComDesDiv 33, was a tall, thin, prematurely graying man in his mid-forties. Gloomy-looking, slow-spoken, and rather taciturn, his habit of reserve was an iron mold from which he seldom broke away. But, like so many men who are hard on the surface, he was human enough when whittled down to the core and the Tragedy at Honda had, indeed, brought out his warmhearted and human side. Throughout the night, Bob Morris had been deeply worried about conditions aboard the Nicholas which, when last sighted, had been on the rocks about 100 yards to the west of the S. P. Lee.

He had kept in steady contact with the Signal Watch maintained on Point Pedernales during the night to keep a lookout for any signal from the stranded destroyer which—with every soul aboard—was on the windward side of the Point and received the full force of biting wind and crushing seas.

Since the destroyer had been constantly enveloped by fog, no signals had been received or sent. In fact, no one could say for sure if the Nicholas was still in existence or if her ship's company was still alive.

When the first faint pre-dawn light filtered through the fog, that still stood opaque and gray along the coast, it found Captain Morris with his signalmen awaiting, with unconcealed anxiety, a glimpse of the Nicholas. He was soon joined by Captain Watson and a large group of officers and men. As the sun rose over Mt. Tranquillon and daylight poured down its seaward slopes, a wind blowing from the land sprang into life. It spread over the mesa, swept out to sea, and carried the fog beyond the shore on its invisible wings. The action was executed unhurriedly and with

dramatic timing. It was as if the jinns of Honda were eager to un-
veil their hellish work with the fullest possible shock effect.

First, the fog receded from the bold and ugly face of Honda
and the sad spectacle of the S. P. Lee came into view. She had
been pushed close under the cliff and had a 20-degree list to
port.

Now, to the left, Bridge Rock emerged from the gray curtain
as did the Delphy. Broken in two the night before, just abaft the
searchlight platform, she was to turn her back on all hopes for
salvage when her forward section took a 45-degree list to port
and rolled slowly onto her side.

Then the jinns swung their awesome wand of vision and
brought the Nicholas into view. Facing seaward, her long slim
hull showed no sign of human life. Great waves broke over her
and sent up huge clouds of spume and spray.

To the experienced eyes of Captains Watson and Morris, it
seemed obvious that the wrecked destroyer sat too deeply in the
water, and with a 20-degree list to starboard, to shelter the crew
within her hull. And yet, it hardly seemed possible that men,
throughout the night, could have endured the waves that constantly scoured her decks with rock-solid seas, and survived. The
signalmen on Point Pedernales waved their red and yellow hand
flags in vain.

No response from the Nicholas. She remained lifeless. Then,
miraculously, a signalman appeared on the port wing of the DD's
bridge.

He semaphored that all aboard were safe. The men on the
Point went wild with jubilation.

They yelled, cheered and shouted as they danced with joy
over this unexpectedly happy turn in the night's tragic events.
Captain Watson and Captain Morris at once set machinery into
motion for the rescue of the men aboard the Nicholas.

Meanwhile, the wall of fog continued its teasingly deliberate
drift to seaward. Now the Chauncey came into view standing upright and high up in its frame of supporting pinnacles. Next, after
a long wait, the heartbreaking sight of the stricken Young on her
beam ends and almost submerged.

Five of them! Five destroyers in the Devil's Jaw! But what
about the rest of the Squadron? Did the remaining DDs escape
unscathed? The answer came, seemingly after endless hours of

unbearable suspense, when the slowly moving fog mass had drifted the entire 500-yard distance from Point Pedernales to Woodbury Rock. First, the wreck of the Woodbury came into view. Next, almost hidden behind her sister-ship's superstructures, the Fuller, her decks awash, appeared. Only a sailor, who has faced a scene like the one that confronted Captain Watson at that moment, could possibly understand or describe the stunning impact of the thoughts and the feelings that stormed through Edward Watson's head and heart at that terrible hour. How is it possible to reconstruct the reactions of a man who suddenly sees the work of his life, his career and his world come to an ignominious and catastrophic end? Not five destroyers on the rocks. Seven! In all the histories of modern navies, there had never been a greater pile-up. But, while the responsibility for the disaster is placed upon the OTC (Officer-in-Tactical Command), credit should be given Captain Watson for the cool unquestioned discipline to which his destroyer men had been trained. Morale of that nature percolates from the top down and not from the bottom up. Were it not for this magnificent discipline on the part of the ship's company aboard every one of the stranded destroyers, the casualties that fatal night could have been enormous. But, bitter as the moment was, this was no time to cry over spilt milk. Like the men aboard the Nicholas, those aboard the Woodbury and the Fuller had to be removed.

If the two DDs had been successful in landing their destroyer men on the adjacent rock, they would have to be taken off. Captain Watson swept the sea with his binoculars for sight of the remaining vessels of DesRon 11. But not one was in the offing. The fog kept its shroud over the sea. Watson telegraphed the following message to the San Francisco Naval Radio Office:

"Woodbury and Fuller struck large exposed rock about 500 yards off shore. Crew of Woodbury on rock, but suffering from exposure. Crew of Fuller still on ship which is listing heavily and pounding on reef. Immediate assistance tug or other vessel absolutely necessary."

At the time of this discovery, the evacuation of the Fuller to Woodbury Rock was just about concluded. In reply to his appeal for help, Captain Watson received a return wire from San Francisco informing him that the Coast Guard Cutter Shawnee, Captain Howell commanding, was already then standing toward the Golden Gate under a full head of steam. The Shawnee's orders were later countermanded.

Crew of the Nicholas Abandon Ship

Even at this early hour, the rocky shores of Honda were cluttered by a tangled mass of wreckage such as bobbing water casks, parts of superstructures, empty rafts, a stove-in lifeboat and one of the S. P. Lee's torpedoes, minus its lethal warhead. More pitiful than this testimony to disaster was the discovery of the body of W. A. Conway, Fireman 1st class, who was flung to his death as he tried to cross the lifeline from the Delphy to the first stage rock.

Lieutenant (jg) Paul Steinhagen, Engineer Officer of the S. P. Lee, was detailed to head a party of men who volunteered to go aboard the abandoned Lee and attempt to pass a stout boat fall, or other heavy line, to the Nicholas by means of the line-throwing gun. Among those selected was Coxswain C. H. Carlson, who played such a major role in establishing the ferry line between the Lee and the forbidding cliff of Honda. Others were: Chief Gunner's Mate E. G. Brown, Chief Quartermaster W. Short, Chief Torpedoman I. Yukovic, Quartermaster Second Class H. A. Bixler, Torpedoman Second Class P. C. Haynes, Machinist's Mate First Class L. M. Goddard, Water Tender Second Class L. M. Woodson, and Radioman Third Class P. M. Kushner.

Aboard the Nicholas, Captain Roesch made ready to abandon ship after a long, wearying, and wave-swept night. "The ship listed heavily to starboard," he recalls, "and she was flooded below. We were afraid that she was about to capsize, so no one went below except for a few moments to get some particular thing. All hands stayed on deck and we rigged life lines to which they could cling. I did not intend to abandon ship before daylight unless she actually broke up or capsized because of the heavy seas washing us and because of the many rocks which could be seen close by. We made ourselves as comfortable as possible and waited for daylight. We did not know where we were and figured we might be near Point Conception. Since we had lost all power, no radio messages were sent out by us."

And yet, contrary to the Captain's expectations, not all hands remained on deck. According to Frederick Fish, Radioman of the Chauncey, whose reportorial instinct had a nose for such tidbits, several men stole below aboard the Nicholas that terrible night. For instance, Gonzales, the wardroom cook, weary of hanging onto his ship while the waves beat him, cautiously climbed into the wardroom, which was filled well up with water, and searched out the Victrola. This he lashed to a steel column and, when the ship was finally abandoned, they found Gonzales bracing himself in a corner, a record playing some ragtime on the Victrola, and Gonzales himself thrumming away on his ukulele.

He was watching the water rise and was ready to "beat it" should the sea rise too high in the wardroom country. Thyne, the ship company's cook, succeeded in starting a fire in the shambles that was once a spotless galley, and made enough coffee to regale himself and a number of shipmates. The galley was at a dangerous angle, and the ship was expected to keel over at any moment.

But most sailors, then as now, would go to hell and back for a mug of java. Lieutenant Steinhagen and his rescue party ferried themselves back to the S. P. Lee aboard one of the life rafts on which they had made shore the night before. The destroyer was lying broadside to Honda cliff and close inshore. Her bow pointed to the narrow sandy beach just north of Point Pedernales and near the spot where the victims of the Santa Rosa disaster had been brought to shore in 1911.

The Nicholas was fast on the rocks at right angles to the S. P. Lee. Her semi-submerged stern pointed toward shore. Because of the relative positions of the two vessels, plus the fact that the only way to get a line to the Nicholas was to shoot it over her bow, it was a tricky shot. Time and again, the light cotton line was carried away by unpredictable eddies of the wind which, constantly freshening, had also increased the action of surf and breakers.

At this time, a Honda character named Tony Cabral projected himself into the rescue picture. Tony was an odd combination of desert rat and beachcomber. While he rented some land from "Don" Sudden, he wasted little time plowing a straight furrow. Instead, Tony spent most of his waking hours hunting or fishing and/or poaching on private preserves. He usually had venison out of season and abalones, too. Senor Cabral was a bear-like creature whose torso carried a thick matting of hair. He could run like a hare, swim like a seal, and swear like a pirate.

Impatient with the slow progress in getting a line to the Nicholas, Tony obtained a coil of light cable from Foreman Tobin's SP stores, tied one end of it around his waist, and nonchalantly set out to make the hundred-yard swim to the Nicholas.

But, even for Tony, the going was too rough on this particular morning. He had to give up the attempt, much to his chagrin and disgust. It did not matter anyway, because, in the meantime, the men aboard the S. P. Lee succeeded in getting a line across where the Nicholas men could reach it.

When the light line was finally secured, one of the Lee's boat falls was attached to it, hauled across, and secured to a stanchion abreast of the after deckhouse. Then a light line was fired from the S. P. Lee to the creek bed's narrow belt of sloping sand. In fairly short order, the boat falls stretched between the Nicholas and a large boulder on the beach.

As Captain Roesch directed the work of rigging the raft that was to serve as a ferry, Fireman Second Class Knox was swept overboard. Catching the action out of a corner of his eye, "Fats" Roesch instantly peeled off his coat and was about to dive in to rescue the man when members of the crew restrained him.

It was some minutes before the Captain could be persuaded that it was too desperate a chance to take. Sturdy, big-hearted man that he was, every soul aboard the Nicholas was of very special value to Captain Roesch.

His usually merry, smiling face seemed to age by years as he watched Knox's desperate battle. At first, Knox was carried shoreward on the shoulders of a swell and was swimming strongly. As the swell changed into surf, Knox was still swimming but seemed unable to escape from being thrown against a rock that was directly in his path. Despite his vigorous efforts, Knox was dashed against the waiting pinnacle and knocked unconscious. Fortunately, quick action by Seaman First Class A. E. Dominick of the S. P. Lee saved the unconscious Knox from being dashed to death on the rocks. Again we are obliged to Editor Adam for an eyewitness story—this time of the abandoning of the stranded Nicholas:

A heavy sea was running by the time the lifeline had been secured between the Nicholas and shore and the work of rescue was extremely hazardous. A life raft or "doughnut," as the sailors called it, was used to ferry the men to shore, six or seven at a time. Crude oil from the bursted tanks of the ships covered the surface of the little bay and the men, as they were pulled ashore, not only received a hard ducking but were covered from head to foot with the slimy oil.

Railroad gangs, and farmers from the nearby ranches, had brought quantities of cable early in the morning and there was sufficient for every purpose in connection with the rescue work. About fifty men were down on the beach to haul in the raft with its human freight every time it was loaded.

All the sailors wore life jackets and although the waves washed over the raft and some were carried overboard all managed to reach shore. As they got near the small strip of beach, they would jump overboard and half swim and wade ashore. The Nicholas appeared to be in a very dangerous position by the time the last load was taken off. The ship had sunk considerably since morning and indicated that it might roll over on its side at any moment.

Most of the deck was submerged and it was difficult for the men to get on the raft. The last load, with the Captain and other officers, had the most exciting voyage of any. Soon after they cut loose, a huge swell tossed the raft as if it were a chip and all on board were thrown into the sea. Some were able to clamber back upon the raft in a few moments but others were carried away from it by the tide. Another swell threw them towards shore, and another back toward the raft so that all were able to get close enough for their companions to pull them on board.

On the mesa atop the cliff above the little beach another bonfire had been built and here the hardiest sailors gathered to warm themselves. Around this camp fire their spirits were as high as boys who had been having a big time in a swimming hole. The officers in their bedraggled gold braid were the only ones who, for a moment, grasped each other's hands and in low tones congratulated each other on their deliverance.

In view of the fact that the sailors had spent the entire night in the upper part of the boat—the hatches having been battened down soon after they struck—and had kept each other warm by huddling close to each other, together with a prospect of almost certain death, the carefree attitude of so many of those boys, when they reached shore, was quite remarkable.

Once on the beach, several men reached the end of their endurance and had to be helped up the cliff where the two Lompoc medicos, Mrs. Atkins, and the pharmacist's mates stood ready to give their much needed assistance as the bruised and limping destroyer men made the hard trek up the steep slope of Point Pedernales. Many collapsed in sheer exhaustion and lay insensible among the scrubby growths of the mesa. Dripping with sea water and black with oil they looked, indeed, like men who had come at close grips with the Grim Reaper.

The Pantry Pirates of Honda

The gang of commissary stewards and cooks, who had assisted "Ma" Atkins in distributing food from John Giorvas' little kitchen, had a problem on their hands. What with the impending arrival of new and hungry contingents from the Nicholas (and perhaps the Woodbury and the Fuller as well) how could they feed those men?

"Well," said one, "if a party can go aboard the S. P. Lee to fire the line-throwing gun, what's wrong with another party going aboard to see what's handy in her galley and storerooms?"

"That would be piracy," said a voluble ship's cook who had a reputation as a sea lawyer.

"Nuts!" answered the author of the idea. "The S. P. Lee is an abandoned wreck. And even if that label doesn't fit—why feed all that good food to the fishes? I'm for hitting the Lee. Anybody who wants to play a game of pirates—follow me."

"Okay, okay," cut in the sea lawyer. "But if we're bent on saving the chow aboard the Lee, what about saving the stuff on the Chauncey? She isn't too hard to get at. What say?"

They all said yes and the newly organized Pantry Pirates of Honda divided themselves into two groups; one made for the S. P. Lee, the other for the Chauncey. But since many of these culinary adventurers were senior ratings, they did not take too much of the law into their own hands, but carefully obtained the consent of the proper officers. The latter agreed whole-heartedly that it was the only immediately available way of meeting the problem of feeding a group that now had swelled to about 500 hungry men.

They knew that Johnny's meager cupboard was as bare as Mother Hubbard's. Not a crumb of bread, not a shred of meat, not a spoonful of coffee remained of the supplies that had come from San Luis Obispo on the night Lark. The Pantry Pirates worked with the speed and skill of veteran scroungers. By the time "Ma" Atkins returned to the Section House with a contingent of Nicholas men from the mesa, a veritable mountain of food was stacked in the Section House. Unofficial figures, reported by Editor Adam, have it that the salvagers collected no less than 15 turkeys, dressed and ready for the oven; 20 hams; great slabs of bacon; cans of ground coffee; tins of butter; sacks filled with bread and other items.

Also, with good forethought, the salvagers had collected a fair supply of pots, pans, and other galley and mess gear. Plus mugs, mugs, mugs for drinking coffee. With the food now on hand, no one would go hungry. The turkeys were soon roasting on well-tended and well-guarded improvised spits over the fires along the track. The hams, fortunately precooked, were made into sandwiches by "Ma" Atkins. She figured that, before she returned about noon to her family in Surf, she had made at least 700 ham sandwiches and hundreds of other varieties. As for bacon—she opined that she had fried so many slices that, if laid end to end on one of the Southern Pacific's rails, they would reach from Honda to Surf. Or, at least, that is the way it seemed that morning to the very tired, very self-sacrificing Angel of Mercy of Honda.

Evacuation of Woodbury Rock

When the fog at long last lifted over Woodbury Rock, the wind was high and breakers threw their weight between every pinnacle and into every crevasse. As the coast line came into view with its seascape of wrecked destroyers, the men on the rock took in the sight with despair in their hearts. It was a sight to bring awe and sorrow to any true seaman. The pride of the Navy had been dealt a solar plexus blow. There lay stranded ships they had known and loved. How many of their buddies had been lost in them?

In the face of this disaster, their own personal sufferings were as nothing. Oddly enough, the curtain of fog was still standing firm to seaward and acted as an effective shield against sighting any squadron mates of the Woodbury and the Fuller that might still be afloat and in the vicinity. Some of the men believed that they had heard the toots of steam whistles through the muffling fog. But those sounds could be from other shipping or they could be destroyer signals.

At any rate, if they were the latter, there was no way of answering them. After recovering a pair of binoculars from the Woodbury and making a close examination of the situation ashore, it became obvious to Captain Davis and Captain Seed that no help could be expected from that direction. However, it was a matter of pressing importance to get the crews off the rock to prevent further suffering from exposure.

The only seaworthy craft at hand was the Fuller's whaleboat, which Ensign Jones and his men had left alongside the Woodbury when they clambered on the rock at dawn. With the seas and surf running as high as they were, taking the whaleboat to shore to find a landing place was a tricky operation. But it had to be tried. Captain Davis took personal charge of the operation. Again Chief Boatswain's Mate Pointer stepped to the fore. Also manning the boat were Radioman Edward Novak, Boatswain's Mate Second Class F. Ross, Coxswain C. O. Lansford, and Torpedoman Third Class Robert Warf.

After some looking around, it was found that difficult, but not too dangerous, landings could be made alongside the Chauncey. Rough as this road to survival was, Captain Davis decided to risk it. Just how back-breaking the operation would have turned out to be will never be known.

Two boatloads of men had been taken off Woodbury Rock to the Chauncey when a large fishing boat came put-putting out of the fog. She was the Bueno Amor de Roma of the Larco Santa Barbara fishing fleet and commanded by Captain Giovanni Noceti. The Roma was to play an important and heroic role in the rescue of the destroyermen who clung so desperately to Woodbury Rock—a task that took plenty of courage, daring seamanship, and long, tiring hours. The story, as it unfolded, was told concisely in the September 10 issue of the Santa Barbara Daily News:

The Roma left Santa Barbara with her sister fishing boat, the North America, at one o'clock Sunday morning to fish off the Honda banks. Neither knew of the disaster. Suddenly, at daybreak Sunday, the Roma found herself trolling past a wreck. The dim outline of a destroyer, the Fuller, could be seen through the veil of fog. Putting in closer Captain Noceti discovered men clinging to the bare rocks.

Forgetting their trolling lines, the fishermen threw over their anchor and rushed for their skiff. They worked until almost noon carrying men from the Fuller and the Woodbury to the Somers and Percival. Their first rescue was of five men who had clung all night long to a projecting reef, unable to reach the main rock because of the terrific seas which were running.

These five men had clung on through the long night hours with the grimness of desperation. But, when they felt the strong arms of the fishermen encircling them, they fell back limp and were lifted aboard by two fishermen, while a third held fast to the rock to keep the frail skiff from being dashed to pieces against the reef.

Those four fishermen worked without ceasing until they, with whaleboats finally sent to their aid by the destroyers standing out to sea, had completed taking the men aboard. This task finished, Captain Noceti ordered his men on with their trolling, in a matter-of-fact way, as though saving men from wrecked vessels and reefs was all in the day's work.

If Commodore Watson had not obtained the name of the fishing smack, the heroic work of the fisherman might have never been traced to Captain Noceti and his men. They are not due to report to Larco's until sometime late Monday afternoon or Tuesday, if fishing has been good, so that the Larco first learned of the part the Roma played in the rescue work from The News today.

"That's Noceti, all over," said Mr. Larco. "He takes each job in hand as it comes along, and undoubtedly does not feel aware of the fact that he and his men have done anything out of the ordinary. The fog must have been heavy, for the Roma and North America fish closely together, and if the North America did not take a hand it was because the fog concealed the wrecks."

"These fishermen did a gallant act," said Squadron Commander Watson.

"They worked through a rough sea, close in upon rocks which, had their boat collided with any of them, would have meant possible drowning of the occupants. The coming of that fishing boat, crawling through the fog, was a cheery sight to the men on those rocks, and I believe recognition should be given the fishermen for their courage and their perseverance."

Because of the persistently prevailing fog, the Farragut, Somers, and Percival did not discover the destroyers stranded on Woodbury Rock until well past 9 o'clock. The Somers was first to break out through the fog bank. She anchored about 800 yards from the Fuller to assist in taking off her crew.

Captain Gaddis also communicated, by whistle signal, with the Percival. She arrived about 10 minutes later and was joined shortly by the Farragut, Flagship of Commander Pye, ComDesDiv 31.

Excepting the few boatloads of men who reached shore by way of the Chauncey's ferry line, the staunch lads of the Woodbury and Fuller were taken aboard the Somers and Percival and later transported to San Diego.

On receiving the news of the pile-up about thirty minutes after it occurred, Rear Admiral Sumner Kittelle, Commander, Destroyers, Battle Fleet, had reversed course of his Flagship, the Melville, and headed for the site of the disaster.

His hopes to make time were foiled by one of the heaviest fogs ever recorded along that fog-ridden region of the California coast. In fact, it was not until Sunday noon, when the weather cleared slightly, that he made out land and breakers. Probing her way through the shifting mists—sounding with hand leads as well as the deep-sea sounding machine—the Melville crept slowly into the Devil's Jaw, foot by careful foot, with Commander Benyaurd B. Wygant, her Captain, at the conn—a term that means directing the handling of a ship and which has been used by seafarers since Noah was an ordinary seaman.

It was a ticklish operation. Even Admiral Kittelle—who wanted to come within visual range of the wrecks—showed a bit of strain as he watched Captain Wygant, who was steady as the rocks that showed their sinister shapes ahead of the slowly steaming Melville. Finally, at 1250, the Flagship anchored in 17 fathoms of water in sight of the stranded vessels as well as the depleted Destroyer Division 31 and the unscathed Destroyer Division 32.

First to report aboard the Melville was Commander Pye, ComDesDiv 31. About an hour later, Commander Roper, ComDesDiv 32, came aboard and made his report. To bring this phase to a close, Divisions 31 and 32 left for San Diego that afternoon around 1500 hours.

Before then, however, Commander Wygant—with a group of Melville officers acting as a Board of Investigation as to the conditions of the stranded vessels—made the rounds of the wrecks. At that time, it was thought that the Chauncey might be saved—a hope that failed to materialize. While he did not lay eyes on the grounded destroyers until noon, Admiral Kittelle had been given a first-hand story, by radio, of what had taken place.

This came about when Captain Watson sent Radioman Frederick Fish back aboard the Chauncey Sunday forenoon to see if his battery-operated radio set was still able to send and receive. If so, Captain Watson wanted to make a report to Admiral Kittelle. It happened that Fish established almost immediate contact with the Melville, whose operator instructed him to stand by. After a short wait, Fish was told that Admiral Kittelle was in the radio shack and anxious to receive an immediate report. When informed how long it would take to get Captain Watson's report, Fish was ordered to give the facts as he knew them.

After pounding his key for the better part of an hour, Fish was finally released and sent in search of Captain Watson, who acquainted Admiral Kittelle with his departure plans and received the Admiral's approval. With the shipwrecked crews provided for, the next step was to get tugs and salvage crews on the job. The Melville, therefore, with ComDes Battle Fleet aboard, got underway for San Diego at 1620, September 9.

The Captains and Their Men Depart

Early that Sunday morning, when it appeared possible that some 300 additional sailors from the Nicholas, the Woodbury, and the Fuller might seek refuge on Honda mesa, Trainmaster Foley, knowing that the two Lompoc medicos were nearing exhaustion from endless hours of uninterrupted effort, sent an SOS to the Santa Barbara Chapter of the Red Cross for medical assistance.

The result was that five Santa Barbara doctors reached Honda in the early forenoon aboard a two-car special train that also carried several newspapermen. The medicos were: Dr. Rexwald Brown, Dr. Hilmar O. Kofoed, Dr. Benjamin Bakewell, Dr. Harry E. Henderson, and Dr. Irving Wills. They took up where Dr. Kelliher and Dr. Heiges left off as they retired to Lompoc with well-earned laurels.

And with the new arrivals from Woodbury Rock, as well as the no less fatigued men off the Nicholas, there was plenty for the newly arrived doctors to do. About noon, another roll call by ship's companies was made. The number of known dead (from the Delphy) still stood at three. The number of missing (from the Young) had been reduced to 20. This because several men, who had drifted ashore above Point Arguello, had showed up, in addition to the five men rescued from a raft. Also present was Boatswain's Mate Braden, who had been rescued by two men on the Woodbury after he had been swept off the side of the Young. Among the injured were 19 seriously disabled and more than 100 ambulatory cases. The special train that had been assembled in San Francisco reached Honda about 3 o'clock.

All its cars had been thoughtfully heated to give added comfort to 38 officers and 517 enlisted men. The special train, which reached San Diego about 3 o'clock Monday morning, made a long stop at Santa Barbara where the Red Cross and the YMCA served a plenitude of coffee, sandwiches, cake, and cigarettes. In Los Angeles, where the train was swamped by an outpouring of reporters and cameramen, Captain Watson refrained from making comment on the accident itself.

However, he paid this tribute to his men: "To command such men is a pleasure and to share danger with them is an honor. To the courage and discipline of American sailors must be given credit for the fact that but 23 men lost their lives in the wreck instead of hundreds."

As one reporter noted: "Hardly a man in the train wore an entire uniform. The costume of many was limited to an undershirt and a pair of trousers, while several still clung to their life belts as an auxiliary covering.

At every station along the route, the wardrobes of the men had been augmented by offerings from sympathetic civilians. One husky sailor strutted forth from the train at Santa Barbara clad in a blanket draped about his shoulders, a woman's hat on his head, bare-footed and with only the remnant of trousers.

"Officers were as badly off as their men, and several who had escaped from their ships clad only in their pajamas, took advantage of the offers of kindly civilians, and wore civilian clothes of motley styles, until they could replace them with more appropriate attire. The effect of the disaster upon the nerves of the men could be noted at every sudden lurch of the train, many of them waking from fitful dozing and leaping to their feet glaring wildly about."

The destroyermen, who for so many hours had played conspicuous roles in one of the world's greatest peacetime naval dramas, had departed from Honda. But their ships—helpless hulks, pounding themselves into scrap iron on the fangs of the Devil's Jaw—would never leave Honda. Seven of them! Seven sleek newly commissioned and battle-ready destroyers that, in minutes, had been reduced to derelicts. As for the hundreds of men who had endured the terrible ordeals of the crashing breakers, the hammering surf, and the sharp rocks, an editorial in the San Diego Union took good measure of their superior stature. In part, it read: The men who faced the terrors of that night are veterans or victims of a battle more terrible than modern warfare waged at sea.

Theirs was a brief engagement in a warfare that will never end until the sea itself shall end, or men unwarmed by a failing sun shall finally and eternally surrender all their ancient hopes of conquering it and faring far. That warfare is more ancient than powder and guns and high explosives and roaring airplanes.

It began with the first frightened and wondering man who ever clung to a floating log. It will end with the twilight of oblivion. Those destroyermen were called upon swiftly and utterly without warning. In one moment there were only waves and darkness. In the next, a sudden roar of breakers, the successive crashing of steel hulls and hidden rocks—then the elemental triumph of swirling waves in the blind maelstrom off the treacherous reef.

There was no time of suspense and preparation, no clearing of the decks, no calls to battle stations, no open and tangible menace of enemy ships and threatening guns. The destroyer men of the Eleventh Squadron had none of the incentive and inspiration that comes in the stern call to naval action in war. Their only ally was their own innate or well-disciplined manhood. That was all—and enough. They met the hour with every ounce of high courage that the hour demanded.

For those bereaved by the brief tragedy that followed, all America must have sympathy and sorrow that is deeply sincere. We can only unite in hoping that the time will come, for them, when the weight of personal sadness may be somewhat lightened and memory shall bring the poignant and profound pride that their loved ones earned in that black battle with the waves.

For the American Navy the tragedy is a victory, though bought, as every victory is, with bitter loss. Seven ships are wiped off the navy roster, but another high tradition has been added to pages already glorious. The destroyer men have shown generations of service men to come, how those who go down to the sea in ships can meet disaster and turn it into victory.

The simplest and shortest summary of this maritime tragedy ever written appears on the Log of Point Arguello Light as recorded by Keeper George Olsen.

In a one-sentence paragraph of truly masterly brevity it reads: "Sat. Sept. 8, 1923—At Point Pedernales or Point Honda on Pacific Coast 3 miles north of Point Arguello on Saturday night, Sept. 8, 1923, 7 U. S. destroyers were wrecked by running on the rocks at 9:04 P.M., all running aground at intervals of two minutes; high seas and a heavy fog."

Thus the curtain fell on the first act of the Tragedy at Honda. Left on the no longer crowded stage was a 16-man Patrol commanded by Lieutenant "Chink" Lee of the Chauncey and Ensign Will Wright of the S. P. Lee.

Their duty was to guard the wrecks and to search for such bodies as might be given up by the sea.

Honda.

An ancient and enduring menace to ships and seafarers from the sixteenth century until this very day. The Spaniards, who plied their tall-masted, high-pooped galleons between Mexico and the Philippines, had a name for Honda. They called it La Guijada del Diablo. The Devil's Jaw.

Today's men of the sea know it as the Graveyard or Cape Horn of the Pacific Generations of sailors have spoken the word Honda with profound respect; thousands of them with soul-gripping fear; hundreds of them in mortal agony.

They still do.

Honda.

STERN VIEW OF USS UTAH (BB-31)

10 - SHIFTING WINDS BLOW GUILT

The Jinns Get the Blame

TREACHERY OF THE SEA AND AIR WERE THE main factors in the wreck of seven of Uncle Sam's sea hornets, destroyers of the Pacific fleet, which pitched at 20 knots an hour onto the jagged rocks at La Honda, 75 miles north of Santa Barbara, Saturday night, according to the belief of naval experts here."

So ran the lead of a news wire story out of Los Angeles filed soon after the Honda Special passed through that city at a late hour on Sunday night, September 9. This angle of approach was, generally, reflected by all news stories of the disaster for the next 48 hours.

On typewriters, telegraph keys, and Mergenthalers clicked colorful and enchanting stories about mysterious current changes along the shores of the Pacific Ocean. This rumor was partially based on a claim advanced by the Captain of the wrecked liner Cuba. He said that a mysterious change in the normal ocean current had pushed him southward and east of his estimated position coming up en route to San Francisco.

Such a current, under the cover of fog, it was pointed out, could easily have pushed DesRon 11 off its course and onto the rocks at Honda. In support of this contention, Captain Knut Knudsen of the steamer Raymond reported, in San Francisco, that he had come up the coast against a wholly unexpected current with a pronounced southeasterly inshore set.

In hunting around for the cause of such a phenomenon, some mariners and oceanographers called attention to the great earthquake staged by Mother Nature in Japan just six days earlier. Almost 40,000 died, about 100,000 were badly injured, and more than 500,000 were made homeless by this terrible disaster. The enormous waves of energy released by the quake could easily, it was explained, have been transmitted across the ocean bottom and set up patterns of wayward oceanic currents. Some scientists held that this was quite possible. Others declared that it was utterly impossible.

Commander L. M. Stewart, Superintendent of the U.S. Hydrographic Bureau in San Francisco, sounded a note of caution when he said that the currents in the region of Honda are always highly variable and that, unless navigators watch their bearings, disaster is likely to follow.

"Eternal watchfulness is the price of survival," said Commander Stewart, "along that dangerous stretch of shore along the California coast."

By September 11, the firms of Honda had completed their run on the front pages of the American press, but they enjoyed a limited engagement abroad.

Now followed a rash of complaints by seafarers about the unreliability of the radio compass—then new and practically untried by the rank and file of mariners. In San Francisco, three old-line steamer Captains voiced complaints about the inaccuracies of radio compass stations along the California coast. Captain John Orsland of the Willamette and Captain Otto Hengst of the Silverado sounded off to their General Manager, Charles L. Wheeler of the McCormick Steamship Co., about confused and inaccurate bearings.

A similar broadside was delivered by Chester N. Gilbert of the Venezuela of the Pacific Mail and a sistership of the wrecked Cuba. The complaints were vigorously refuted in a newspaper interview by Lieutenant J. M. Lewis, who had just finished a tour of duty as radio materiel officer of the Thirteenth Naval District. He spoke straight from the shoulder and declared that nearly all of the criticism of the Radio Compass Service came from a bunch of barnacle-covered shellbacks.

"The attitude of these old-timers," he said, "seems to be that any new navigational invention or idea is no good and that it should be scuttled immediately. They hold to the course that no landborne radio operator is going to tell them where their own ships are. Yet, when they get in a pinch, they call the stations. Being prejudiced against them, they fail to cooperate and claim that the bearings they get are wrong."

Meanwhile, not a word of explanation came from Admiral Samuel S. Robison, Commander-in-Chief, Battle Fleet, or from the Navy Department in Washington. Again speculation as to the part the stranding of the Cuba had played in staging the disaster swung into the foreground. Representative of this badly inaccurate forward pass among the Monday morning quarter-backs is the following news excerpt:

"Because of the wreck of the Pacific Mail liner Cuba on San Miguel Island, off the coast nearly opposite the place where the destroyers met their doom, the radio operators on shore were trying to direct the course of the destroyer Reno to her assistance. The fog was so dense that the radio was the only guide.

"The operators of the destroyer squadron, speeding along in the fog over a rough sea, picked up the signals intended for the Reno. The mistake was discovered and Arguello Radio Compass Station corrected the bearings for the destroyer fleet. The correction came about five minutes before the wreck. The ships changed their course but the correction came too late."

The morning papers of September 11 carried the following Associated Press story out of Washington based upon attitudes and opinions within the Navy Department:

Sept. 10—Lacking even the barest official explanation of the loss of seven first-class destroyers on the California coast, Navy officials tonight continued to withhold judgment on what they termed the most severe peace-time blow the Navy has ever sustained.

Although regulations prescribe that every effort be made to forward immediately names of dead and injured in such cases, no such list has been received at the department up to a late hour tonight.

The initial dispatch from Admiral Robison, commanding the Pacific fleet, informed the Department that specific orders had been issued for the preparation and relay of this list, the duty being assigned to Captain Edward H. Watson, commanding the wrecked squadron.

Theory advanced in press dispatches that radio operators on the destroyers were thrown off their reckoning by shore signals intended for the Reno was declared by officers in the department to be doubtful.

Leaving out of all consideration, they said, the material difference in location of the wrecked mail steamer and the destroyer group, "position signals" invariably are addressed specifically to the ship which has requested them. This was held to render it improbable that all of the operators on the naval vessels could have taken the signals as bearing upon their own course and to have acted in concert, although one might have done so.

Unofficial description of the scene of the wreck and known peculiarities of the coastal area in which it occurred, led to the belief by some officials that a tidal disturbance of unusual force threw the destroyers far off their course, probably without the knowledge of the officers on board. A possible connection between such a phenomenon and the recent Japanese earthquake was discussed. Records of the hydrographic office and reports of naval officers who have served extensively on the California coast agreed that the Santa Barbara section frequently experiences a coastward tide, attributable to no known marine factor.

Secretary Denby said the department would await the report of the board of investigation, organized automatically as soon as the accident was reported to the Commander in Chief, before he would decide whether an inquiry would be ordered to fix personal responsibility.

In the meantime, he asked that the public follow the department in withholding its judgment. With this summary, the avid public interest in the disaster had to remain satisfied. There was nothing to do but wait. However, aroused public sentiment waits no more than do Time and Tide.

Salvage Operations at Honda

The funereal silence that usually spreads its brooding weight over the waters and shores of Honda was not restored for many weeks after that fatal moment when the Cavalry of the Sea galloped into the Devil's Jaw and came to a world-shaking halt.

There was not only the 16-man Wreck Patrol, commanded by Lieutenant C. V. "Chink" Lee of the Chauncey, assisted by Ensign William "Will" D. Wright of the S. P. Lee, but also two score or more members of salvage crews from minesweepers and Navy tugs.

These latter had the dangerous and difficult job of salvaging torpedoes, deck-guns, radio sets, confidential papers and other equipment within human reach. It was soon established by Captain D. C. Nutting, of the Navy's Construction Corps, that none of the seven stranded ships could be got off the beach.

However, much valuable equipment could be and was recovered. He urged that the salvagers should work at all possible speed to complete the job before the heavy seasonal winds, that begin to blow in October, would put an end to it. Captain Nutting estimated that the monetary loss to the United States caused by the disaster would total $13,500,000.

During the early days of the salvaging operations, Rear Admiral Kittelle, in the Melville, directed operations from offshore. Since it was generally believed that some of the Young's missing crewmen had gone to their deaths in their ship, men with cutting torches boarded the capsized Young when the seas permitted and burned holes in her port side abreast the living and machinery spaces. Despite a careful search, no bodies were found. Thus proving that the missing men were swept into the sea.

The salvagers were in luck when it came to recovering the 80-odd torpedoes aboard the wrecked destroyers. Whenever possible, the "tin-fish" were fired from the deck torpedo tubes of the stranded ships—in which they had remained loaded—out to sea for recovery by the minesweepers. Otherwise, they were pulled out by teams or tractor from the shore.

Afterward, these heavy and expensive engines of destruction—each weighed about a ton and cost about $5000—were pulled up the cliff, placed on sleds, hauled to the railroad track, and swayed into railroad freight cars.

To paraphrase Mr. Milton's maxim—they also serve who only stand and watch. That certainly was true in the case of the 2 officers and 16 men whose dreary chore it was to be detailed to one of the most unpleasant, exhaustive, and nerve-frazzling jobs that arose out of the Tragedy at Honda.

During the daytime, they searched for and, now and then, recovered the badly mangled and decomposed bodies of dead comrades. At night, heavily armed, they patrolled the cliffs along the water for prowlers who sought to loot the wrecks and salvage whatever they could lay hands on.

In spite of their vigilance, looting did take place, probably by boats at night. Stories were current of Navy furniture, silver and tableware that found its way into the black market ashore. But that was not all. From early forenoons to late afternoons they would be called upon to act as guides and traffic officers for the thousands of visitors who swarmed to Honda mesa day upon day.

From north and south, spectators came by early morning trains. To north and south they departed on late afternoon trains. The Sudden Grade—the only road to the mesa—was made unusable for auto traffic early in the afternoon of September 9. Only a wagon road, it soon became so deeply rutted by scores of cars that constables had to be placed on traffic duty to turn the vehicles back before they all bogged down.

While the inrush lasted, Honda took on the unbecoming atmosphere of a combination picnic ground and carnival. The goggle-eyed spectators did not come to view the sad settings of a great naval tragedy in moods of veneration but to raise crops of gooseflesh and otherwise to have a good time.

Hawkers did a land-office business selling postal cards with photographs of the stranded destroyers, and shady operators did a brisk under-the-counter trade in more or less authentic souvenirs. Barkers offered tepid cold drinks and lukewarm hot-dogs. What with the seven destroyer wrecks lying offshore, like tombstones over watery graves, the scene was much like that of a county fair in a cemetery. Worst of it all, so far as the patrol was concerned—the newspapers called it the Death Watch—was that it was left without food or shelter when their fellow destroyermen boarded the special train for San Diego.

As sailor's luck would have it, the seas were so high during the first few days after the disaster that no boats from the Melville, or other ships standing offshore, could land provisions and other essential items. A fairly realistic picture of what this Death Watch had to endure was drawn, editorially, by the *Santa Barbara Daily News* on September 11:

Castaways at Honda

Apparently there is a loose cog in the United States Navy's machinery. The seven destroyers which ran ashore in the fog at Honda struck Saturday night. Those on board escaped with scanty clothing and only such equipment as they could carry with them, which was negligible. Most of the survivors were shipped promptly to San Diego where they could be cared for and re-outfitted.

A detail of 16 men and 2 officers was left at the scene of the wreck to watch for bodies and to prevent looting of the battered ships.

Nearly three days have elapsed since the disaster and yet those men are still camped like refugees on the windswept bluff without shelter and with inadequate supplies.

A car load of supplies has been sent to these sailors by the Santa Barbara Red Cross to furnish Navy men on duty with the things that they should have had from the Navy Department.

Although these sailors are camped along the main line of the railroad, within a few miles of markets where can be obtained everything they require, they are still, to all intents and purposes, in the condition of castaways.

These men are officers and men of the United States Navy on official duty just as much as though they were on their ships. To them has been assigned the duty of guarding millions of dollars' worth of property of the government as well as the sad task of seeking and caring for the bodies of their dead comrades.

The only available building at the scene of the disaster has been converted into a morgue. There, the dead are housed. The living sleep in the open, on a wind and fog swept bluff, with few blankets and without any camp equipment. Kindly disposed women and sightseers have brought the sailors cakes and cookies and doughnuts. Leftovers from the lunches of the tourists— who have come by the hundreds to see warships that cost from $10, 000,000 to $15,000,000 lying in surf as heaps of junk— have gladdened the hearts of men engaged in doing the nation's work.

The United States Navy is a remarkable organization. It has a record of which the American people are proud. But it is difficult to understand why the conditions at Honda are as they are. It surely takes a long time for Uncle Sam to unwind the red tape in which his Navy Department is bound. The points of complaint voiced in this well-meant but rather premature criticism were met within a few days when the seas subsided.

As soon as it was safe to lay boats alongside the starboard side of the Chauncey, provisions and supplies of all kinds were sent ashore from the ample stores of Navy vessels offshore, among which were the Melville, Undaunted, Tillamook, Tern, Ortolan, and the hospital ship Relief. For the remainder of their duration on duty at Honda, the men of the Wreck Patrol fared reasonably well. Their only grievance, if they had any—which is doubtful—was that, unlike their comrades, they did not make an early return to San Diego and their homes, plus brief furloughs. Like their ship and squadron mates, they had endured and won hard-fought battles for survival from shipwreck.

But, unlike them, they had to stand to for additional duty when the others had none. For the fine execution of those unpleasant and arduous tasks the men of the Honda Patrol, whose names follow, deserve a hearty "Well done!"

Lieutenant C. V. Lee, Senior Patrol Officer; Ensign W. D. Wright, Junior Patrol Officer; A. Van Warren, Chief Torpedoman; R. M. Reynolds, Chief Pharmacist's Mate; H. E. Goger, Water Tender 2nd class; D. G. Jackson, Engineman 1st class; F. W. Fish, Radioman 1st class; C. J. Spaulding, Chief Quartermaster; and G. J. Burke, Yeoman 1st class. R. J. Wutz, Boatswain's Mate 1st class; C. A. Russel, Ship's Carpenter 2nd class; H. Fels, Engineman 1st class; P. M. Kushner, Radioman 2nd class; E. Novak, Radioman 3rd class; J. O. O'Reilly, Chief Quartermaster; T. Avery, Fireman 1st class; G. H. Cochran, Boatswain's Mate 2nd class; and S. V. Sage, Seaman 1st class.

As will be observed, Frederick Fish, the writing radioman of the Chauncey, was a member of the patrol. His comment—made at the time and therefore hot off the griddle—is picturesque.

"This patrol, with shattered nerves and bodies weak from loss of sleep and insufficient food, performed the hardest of tasks for a period of many days," he said.

"With but 16 men they guarded the wrecks night and day, to keep off looters; patrolled the beach from dawn to dark for bodies; worked like giants to salvage material from the wrecked ships, and served as pack horses to lug bodies and materials from the wrecks to the Railroad Station and Section House at Honda. These duties were performed under the hardest of conditions, irregular and scanty diet, sleeping quarters on the ground where the blankets were soaked through by the continued fogs and drizzles. Most of the patrol obtained but 2 or 3 hours sleep a night, and many became so exhausted that they could not sleep when opportunity offered.

"Telegrams and phone calls poured in from anxious families, and each was treated with utmost consideration; information being given immediately, where available. The press was given all the information possible that would tend to alleviate the load of doubt and grief that held the relatives of the missing men in its grip. On the fourth day of their grim watch, loads of cigarettes, food stuffs and blankets poured in on the patrol. The Navy had not as yet had an opportunity to get supplies to the patrol, and citizens from Lompoc, Santa Barbara and San Luis Obispo did more than their bit in providing for the men and making them comfortable. Too much credit cannot be given to the good folk of the surrounding towns for their generous response to the patrol's need.

"During the following days of their grim watch for dead bodies, the naval patrol was dubbed the 'Death Watch' by the newspapers. The patrol succeeded in finding 12 bodies and salvaged much material from the wrecked destroyers."

By October 17—long after the Honda Patrol was withdrawn—five more bodies had been recovered. This brought the total to 17.

The remains of the six men unaccounted for were never found. Soon the salvagers and their ships departed. The civilian firm of Chapman and Scott had been awarded a salvage contract. Soon the stream of visitors died to a trickle and then faded entirely out. Soon Honda mesa was its old forbidding, desolate, and inhospitable self—barren and gloomy like the lands God gave to Cain. Day on day, week on week, the pounding wave action and the grinding of the rocks turned the stranded destroyers into smashed and twisted skeletons. The waves picked the wrecks bare as bones, the rocks chewed them into rusty remnants.

Bit by bit, the heavy swells and crashing breakers pressed the dead destroyers into deeper water. At long last, the Devil's Jaw, after crunching the last morsel, settled down to wait—with the deadly patience of the infinite—for the jinns to provide another feast.

Secret Inquiry Backfires

Up to now, not a shadow of public criticism had fallen upon—nor a finger of guilt pointed at—Captain Watson and his destroyer captains in DesRon 11. In fact, if anything, Captain Watson had been a figure marked for widespread public sympathy. To be sure, public opinion was a heaving sea of intense and voluble interest, but so far the favorite sport of crusty old shellbacks and destructive critics—to blast away at new-fangled electronics—had been both absorbing and rewarding. Why stoop to the obvious of picking upon mere naval officers when more fascinating pursuits were in the offing?

Thus it was—when Admiral Robison, on September 11, appointed a Court of Inquiry—that many newspapers in various parts of the country hastily told their readers—as did the San Francisco Chronicle—that while they "deplored the naval disaster, the mishap itself does not necessarily mean that somebody was to blame."

Because this extremely well-written editorial deals understandingly with several prominent aspects of the Tragedy at Honda, it is presented here in toto: Naval regulations provide for the convening of such a Court on occurrence of any matter deemed serious enough to require thorough investigation. No officer of the ill-fated destroyer squadron is on trial. The proceedings of a Court of Inquiry are in no sense a trial of an issue or of an accused person. The functions of a Court of Inquiry are to sift facts for the information of the convening authority and to assist him in the performance of his administrative duties. Such a Court does not relieve the convening authority of responsibility for his administrative acts. Naval instructions and procedure provide that the Court of Inquiry shall go thoroughly into the matter and include in its findings a full statement of the facts deemed to have been established. The Court is further instructed to give its opinion as to whether or not any offenses have been committed or serious blame incurred.

In other words, a Court of Inquiry affords the best means of collecting, sifting and methodizing information for the purpose of enabling the convening authority to decide upon the possible necessity or expediency of further proceedings. For the public to assume, without inquiry and in the absence of adequate understanding, that someone necessarily was guilty of culpable negligence, rashness or inefficiency which resulted in the recent disaster, would be manifestly unfair. Possibly someone was at fault, but not necessarily so.

Captain Stanford E. Moses, assistant commandant of the 12th Naval District, commenting recently on what to the layman might appear to be unnecessary risks, said: "It is a fundamental truth that you cannot train men for war in times of peace unless during that training they accustom themselves to risks which might arise in war."

The late Admiral Togo—famous for his victory in 1905 over the Russian Fleet at Tsushima—once said: "The life of a naval man is a never-ceasing war and whether the country is engaged in war or not makes no difference in his responsibilities. In war he may display his strength—in peace he should acquire it."

We believe these views to be fundamentally sound. Moreover, while every loyal American is bowed with grief at the sacrifice of precious lives, and while we deplore the loss of our ships, it is well to bear in mind that since our first destroyer was put in commission, twenty years ago, American destroyers have cruised more miles than have the combined destroyers of all other nations, and with comparatively few mishaps. We do not condone rashness in times of peace, but neither can we advocate timidity.

The Court of Inquiry named by Admiral Robison consisted of Rear Admiral William V. Pratt, Commander Battleship Division 4; Captain George C. Day, Commander Submarine Divisions, Pacific, and Captain David F. Sellers, Commanding Officer, Naval Training Station, San Diego. Admiral Robison also requested Rear Admiral Sumner Kittelle to appoint a Judge Advocate from his Destroyer Squadrons. Lieutenant Commander Leslie E. Bratton, Captain of the Stoddert of DesDiv 32, was named to that post. Lieutenant Hardy B. Page, also of DesDiv 32, was appointed Assistant Judge Advocate.

Rear Admiral Pratt was a former Assistant Chief of Naval Operations.

Captain Day had served as Fleet Navigator of the Great White Fleet which President Teddy Roosevelt sent around the world in 1908. Captain Sellers, later to become Judge Advocate General of the Navy, was a distinguished battleship commander. Bratton, besides being an active destroyer man, was a graduate lawyer and well versed in Admiralty procedures.

The Court of Inquiry, over which Rear Admiral Pratt presided as Senior Member, convened in a large conference room in the Administrative Building of the Naval Air Station on North Island, San Diego, on the morning of September 13. An executive session, it was not open to press or public. When news and cameramen converged upon the building to gain admittance, they found themselves face to face with a squad of marines.

After marking time for the better part of an hour, the newsmen obtained a most unsatisfactory interview with Admiral Pratt. His remarks, when screened to their absolute essence, were to the effect that the Inquiry would be conducted in secrecy behind closed doors. However, the findings, in time, would be made public when the records had completed the red-tape circuit of Navy channels. That might be a matter of weeks, if not months. For news-hungry reporters trying to fill the demands of a news-hungry public this was indeed a bitter pill. But they did not swallow it.

Instead, they crushed it with angry indignation. Star chamber—secret session stories covered the front pages of newspapers that were spewed forth that afternoon by the roaring presses of the nation. Protests against secrecy rolled toward Washington from every part of the country with the force and fervor of overlapping tidal waves. What was there to hide? The public wanted to know. Bring all facts out into the open!—the press demanded.

Secretary Denby; Admiral E. W. Eberle, his Chief of Naval Operations; Admiral Robert E. Coontz, Commander-in-Chief of the United States Fleet; and others went into executive session. Admiral Coontz was embarrassed and angry because, up to then, he had not received, through channels, a satisfactory explanation from Captain Watson of the Honda disaster. After the meeting, two things happened:

Admiral Coontz announced that the Court of Inquiry would be instructed to get to the bottom of the catastrophe and that no one who was guilty would escape. Next, late evening news from San Diego carried the surprising and upsetting disclosure that the total of destroyers grounded at Honda was nine, not seven, as previously reported.

Sharp-nosed newshounds on the West Coast were baying the tidings that the destroyers Somers and Farragut had been placed on the binnacle list by virtue of the damages they incurred as they backed away from the reefs they barely touched that fatal Saturday night. Once more, Fate had intervened so that somebody fumbled the ball and built up unwarranted pressures of suspense and suspicion.

In some unexplainable manner, according to the Secretary of the Navy, as quoted by the press, the report of the Inspector Pacific on the Somers and the Farragut had been sent to the Board of Inspection and Survey in the Navy Department when it should have been sent via the Division of Communications.

Actually, this was a matter of small import, magnified out of all proportion by the tension under which the Navy Department, as well as ComDesRons, Battle Fleet, and his subordinates were working.

True, the reports of grounding of the Farragut and the Somers could properly have been sent by dispatch, but Captain Watson showed commendable foresight in having these damaged ships inspected—an inspection which had to be done by divers—before he reported the extent of their injuries.

It is entirely understandable that the Chief of Naval Operations must have up-to-the-minute information regarding the availability of all ships.

However, the Farragut and the Somers were not badly damaged and, considering the mass of administrative duties—most of them related to personnel and human relations—resulting from the Honda disaster, under which Captain Watson, his subordinates, and his seniors, were laboring, a little more consideration and latitude would have been especially appropriate.

What was really needed to clear up the confusion existing in the minds of the press and the public was a clear statement, from top levels, of the entire situation.

First five destroyers had gone on the rocks. Then the number was found to be seven. Finally, the total had grown to nine. All this could have been explained easily by the proper people. Likewise, it could have been pointed out that the real cause of the disaster could not be determined until a Court of Inquiry had gone into all the facts.

In fairness to all concerned, the findings of this tribunal would have to be awaited with patience and, it was to be hoped, with forbearance. The report of Inspector Pacific, as published in the morning newspapers of September 15, reads innocently enough, considering the role it played as a time fuse to blow up standard procedures by Naval Courts of Inquiry.

It follows:

To Board of Inspection and Survey:

Completed inspection of /. F. Burns on Sept. 11.
Requires immediate renewal port forward main turbine gland.
Repairs to generator armatures. (This ship was not concerned in the disaster.)
Completed inspection of Farragut and Somers Sept. 12. Owing grounding on Sept. 8 Farragut damaged as follows:
Saltwater leaks into C-101, reserve feed tank, requiring make up feed direct from evaporators. Saltwater leaks into fuel oil tanks B-109, B-111. Starboard propeller one blade damaged. Bottom dented.
Frames slightly buckled at frames 101, 102 and 109 starboard side. Topside dented near frame 90 by collision.
Somers damaged by grounding Sept. 8 as follows: A-101 hole, crescent shaped, 5 inches by 14 inches. A-102, A-104, A-105 saltwater leaks. Bottom dented. Frames buckled near stern. Propeller damaged, causing heavy vibration.
Recommend Farragut, Somers be docked San Diego examination and emergency repairs sufficient proceed navy yard, Mare Island.
(Signed) Inspector Pacific.

According to a United Press dispatch from Washington on the morning of September 15, Secretary Denby issued the following statement in a Navy Department press release:

"I have ordered that the fullest publicity be given to the investigation to be conducted by the Court of Inquiry. This is a very unusual course, and one entirely contrary to custom. But I have taken this step because of the mysterious circumstances surrounding the wreckings and the sensational rumors that since have arisen."

Just how deeply this deplorable foul-up cut into the official and personal pride of President Coolidge's Secretary of the Navy was revealed as follows by the Army Air Navy Journal:

"Secretary Denby expressed himself rather strongly concerning the extraordinary delay in sending reports to the Navy

Department on the wreck of the destroyers. He was especially incensed over the delay in sending to the Department a report on the injuries to the destroyers Farragut and Somers not mentioned in the wreck of the seven others.

"The Secretary feels that he has been placed in the attitude of attempting to suppress the news of the wreck. The same may be said, too, of high-ranking officers in the Navy, whom it was assumed by the representatives of the press should have been advised as to the character of the disaster."

Overnight, speculations on the potential human elements of error that might have caused the Tragedy at Honda became pressure cookers that were filled to the exploding point with all sorts of wild guesses and vicious gossip. The only way to prevent a blow-up was through quick use of the safety valve of complete publicity. Now the shadows fell on, and the fingers pointed at, Captain Watson and his Destroyer Squadron. As suddenly as a tropical sunset, the bright picture of courage, sacrifice, and discipline—revealed in the rescue of more than 700 destroyer men from their wrecked vessels—was replaced by black-as-night rumors of incompetency, recklessness, and insobriety.

No fiction produced by over-imaginative minds was so low or so spurious that it could not bear mouth-to-mouth repetition. This was particularly true with respect to insidious tales about the use of liquor in the Navy.

In 1923, Prohibition was a leading religious, political, and criminal issue. There was a generous sprinkling throughout the land of irresponsible zealots who would not hesitate to smear anyone with the taint of bootleg alcohol.

Some of these so-called "good" people launched an avalanche of scandalous gossip about insobriety aboard the destroyers. Many of their home-distilled tales found their way into print. At this point, it may be well to say that implied violations, with respect to the presence and use of alcoholic spirits aboard any ship in DesRon 11, were thoroughly refuted during the Court of Inquiry proceedings. Today most people have forgotten that no liquor had been permitted aboard ships of the U.S. Navy—other than the doctor's "medical red"—for 9 years prior to the Honda Disaster.

By General Order 99 on June 30, 1914, Secretary of the Navy Josephus Daniels had abolished the wine messes aboard all naval vessels.

Such messes had been permitted since the earliest days of the Navy—and still are permitted in most of the navies of the world. But, with the passing of the age of sail, ships no longer spent weeks—and even months—at sea without touching at a port.

The Powers That Were in Washington, therefore, decided that the ancient custom of "splicing the main brace" could be dispensed with. As a song of that day had it:

Bryan he drinks grape juice,
And Daniels likes it well.
They'll make the Navy drink it—
They will like hell!

But they did. And, after the first natural resentment wore off, the Navy found that the new order was really a relief. Even during Prohibition, which began in 1920, there was liquor to be found ashore—and alcoholism was never a problem in the Navy. Men inclined to look too intently upon the wine when it was red eventually found themselves "on the outside," where they could not be a menace to their shipmates or to ships which each year were becoming more and more a mass of intricate machinery and equipment requiring the most sober and expert handling.

The penalties for violating the Navy Regulations by drinking aboard ship were too severe and, because of close quarters, the chances of detection were too great, to encourage even the most reckless to take a chance.

When Captain Watson's official report on the wreck was finally written and received in Washington, it did not contain as much information about the disaster, and the circumstances that surrounded it, as many observers had been hoping. It was nothing more than a concise resume of facts already known and fully publicized. It read:

At about 8:50 P.M., on September 8, 1923, after checking positions on chart, the squadron commander decided to change course to 95 degrees at 9 P.M. to make the approach to the Santa

Barbara Channel, and gave orders to that effect. The course was changed as ordered and at 9:05 the Delphy stranded on Pedernales Point, known locally as Point Honda. The squadron commander immediately sent warning signals to the vessels astern, but due to the configuration of the coastline, the outlying rocks, and the fact that the ships were in formation, the following vessels were unable to avoid the danger, and were also stranded: S. P. Lee, Young, Woodbury, Nicholas, Fuller and Chauncey.

Although seven ships were suddenly stranded on a rocky promontory on breakers at night, and the lives of 800 officers and men were immediately imperilled, only 23 men are missing.

This regrettable loss of life was incredibly small under the circumstances, and can only be ascribed to the mercy of Providence and to the high state of discipline and morale of the personnel of the squadron, worthy of the highest naval traditions.

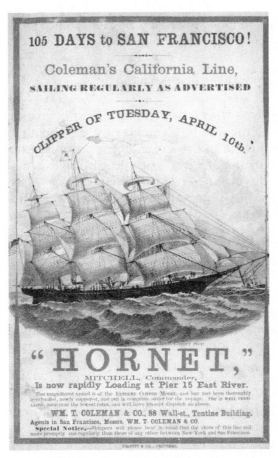

AD FOR THE HORNET CLIPPER SHIP OF THE 1850'S

11 - Echoes of Pile-Up Roll Like Thunder

Wheels of Justice Start to Grind

IF THE CROWDS THAT CAME TO GAWK AT THE wrecks off Honda behaved like picnickers in a cemetery, then the air that pervaded the conference room of the Court of Inquiry had a touch of the gruesome and calloused curiosity that hung over London's Tyburn gallows during the era of public hangings. This flavor was mainly imparted by representatives of the more sensational elements of the press.

In 1923, the tabloids were in their lustiest diaper days and used the boldest, blackest, and biggest headlines to expand the beach-head they had established on the circulation fringe of the nation's newspapers. Hardest hit by these assaults were the so-called yellow sheets. To hold their ground, they fired equally heavy broadsides of headlines. And, what with the related elements of danger, death, and rescue at sea, the fixing of the blame for the Tragedy at Honda was a made-to-order target for their crossfire.

This pressure toward sensationalism did not, of course, affect the members of the Court of Inquiry. Nor was it reflected in the treatment accorded the proceedings by the vast majority of conventional newspapers. But it did have an impact on public opinion and it did add to the strain of uneasiness to which witnesses and defendants in the action were subjected.

In the day-to-day inquiry by the Court, the scandalmongers had rather poor pickings. The trial was wholly void of the double-edged, aggressively accusing questions traditionally bellowed by politically minded Public Prosecutors to whom courtrooms are filling stations at which they service their ambitions.

Lieutenant Commander Bratton, the Judge Advocate of the proceedings, was a soft-spoken, quiet-mannered man who did not engage in histrionics. A lawyer well versed in admiralty law, and an experienced destroyerman, he had complete knowledge of the factors involved. Bratton's task; to prosecute the case against his superior officers and squadron mates—added to his burden of unpleasant responsibility. But he did not let friendships stand in his way. There is a vast difference between a civil and a military court. In the former, defendants who know the judge or justices personally are relatively few.

In a military tribunal, just the reverse is the case. In most instances, senior officers who appear before Courts have either served with or under the men who sit in judgment. Therefore, in addition to the suspense and dread that form part of the atmosphere of any trial, a Naval Court contains a flavor of almost implacably impersonal lack of partiality.

The Court of Inquiry began its 19 days of questioning defendants and witnesses on September 17. During this time, San Diego was an important news center of the country. The sessions were held in the conference room in the Administration Building of the Naval Air Station. This spacious chamber was packed to the bulging point by members of the Court, parties to the proceedings, newspapermen, and spectators.

In line with the informal judiciary status of a Court of Inquiry, the room was not arranged with the trappings of a Court Martial. Captain Day, Rear Admiral Pratt, and Captain Sellers in that order from left to right, took their seats along the window side of a huge table that ran along the wide front wall of the room.

At one end of this table, the prosecution—Lieutenant Commander Bratton and Lieutenant Page—had their places. In a chair, across the table from and facing the Court, sat the witnesses as they were summoned. At other tables were places for the 12 defendants and their counsel, as well as representatives of the press. As Senior Member of the Court, the burden of its conduct fell upon the capable shoulders of Rear Admiral Pratt. He was a man of medium stature. Years at sea had left their etchings upon his craggy features. His heavy eyebrows spread like the wingspan of a stormy petrel over eyes that were lighted by keen intelligence and stern determination. As an inquisitor, he had a quick and biting way of cracking through the shell of testimony to get to the kernel of evidence. The first action of the Court was to proceed, with Captain Watson, Captain Morris, and Commander Pye, as well as the Captains of the stranded vessels, to the Destroyer Base. There the crews of the wrecked ships were quartered in ship's companies pending future assignments. The crews were mustered before the Court. After each commander had read his report—mainly technical—the men were asked if they had any comments, complaints or additional information.

None was offered.

Before the groups were dismissed, Captain Watson—gray-faced and in a voice that held a dull and weary ache—paid this tribute to the enlisted men of the wrecked destroyers: "Although preoccupied with observing the movement of the ships following the crash, I noticed with satisfaction and surprise the coolness and courage of the young enlisted men. They remained at their stations. The men on the bridge of the Delphy were perfectly calm and self-possessed, and went slowly aft when told to abandon the ship by the Captain.

"There were fourteen ships in column at 150 yards distance between stern and bow. The grounding took place suddenly in the dark, in heavy surf on a lee shore and at night. The crews were largely—more so than ever before in the history of the Destroyer Service—made up of young and only partially trained men. Cognizant of this fact, and observing the strandings of the ships, I was of the opinion that the catastrophe could conservatively cost the lives of 200 to 300 men. Owing to the fact that the ships were in close formation and to the configuration of the coast line and outlying rocks, a disaster of this magnitude seemed inevitable.

"The regrettable loss of twenty-three men is the cause of deep distress and soul-searching. Nevertheless, it was incredibly small, and can only be ascribed to the mercy of Providence and the high state of discipline in the personnel of the Squadron, worthy of the highest tradition of the Navy."

Blodgett Caught in a Trap

When the hearings settled down to their normal stride, it immediately became apparent that the initial tug-o'-war between the opposing forces of defense and prosecution would be fought over the question of the reliability of the bearings received during the early evening hours aboard the Delphy from the Point Arguello Radio Compass Station.

And the point of contention was whether or not the Compass Station had received and complied with a bearing request from the Delphy at 2035 hours: a short half-hour before the pile-up. Previous bearings provided by the station on that fatal evening indicated that the Delphy was to the north of Point Arguello. But Lieutenant Commander Hunter, her skipper, and Lieutenant Blodgett, Executive Officer and Navigator, believed that she was south of the Point. In response to a message from the Delphy saying: "We are to the southward; give us a reciprocal," the Compass Station, at 2035, was alleged to have replied: "You bear 168 degrees, true, from us."

Such a bearing, if accurate, would definitely—as earlier explained—have placed the Delphy and the other destroyers to the south of Point Arguello. This was the crucial point, and on it the navigational defense stood or fell.

During the first days of the trial, Lieutenant Blodgett was called to the witness stand. He had been Executive Officer on the Delphy since 1920.

In February 1923, he had won his promotion to Lieutenant (jg). The essence of his testimony was that he shared Lieutenant Commander Hunters views in not taking the bearings given that night by the Point Arguello Compass Station too seriously. He described how their dead reckoning led them to believe that the Squadron was south and not north of Point Arguello.

In contrast, the bearings sent by the station were confusing. "They kept giving our position north and to the west of Point Arguello," Blodgett explained, "and when we could not make this check with our figures, we finally took the reciprocal of their bearings, which would show us already in Santa Barbara Channel."

At this point, Admiral Pratt leaned forward across the table, his craggy brows contracted in a foul-weather frown.

"Mister Blodgett!" he exclaimed in a voice that had the brake-power of reversed propellers. Blodgett stopped his flow of words.

"Before you go any further," continued the Court's President without a pause, "it is necessary to inform you that you appear to be involved in this inquiry and that it is my duty to name you as a defendant in this case."

Actually, there was nothing surprising about Admiral Pratt's action. In Blodgett's position aboard the Delphy, it was quite to be expected that he would be named as a defendant.

However, the sensation seekers at the press table stopped doodling and began scribbling. He was grist for their mills. Another lamb led to the slaughter. Now the total of defendants stood at 13. Just how unlucky would that be for those whose very careers were at stake? The swift change of Lieutenant Blodgett from the status of a witness to that of a defendant made the officers whose acts were questioned look at each other with glances that reflected unhappy speculation. Up to now, most of them had been of the belief that they would be shielded by the follow-the-leader concepts of the destroyer doctrine.

They recalled that, at the time he gave testimony, Lieutenant Commander Laurence "Jasper" Wild, Communications Officer of the Squadron, had made it plain that only one destroyer, the Delphy, was permitted to ask compass stations for bearings. All the others must follow in the flagship's wake. He had added that destroyers in the Squadron would have violated orders if their radiomen had left the Destroyers, Battle Fleet, and the DesRon 11 bands unprotected long enough to ask for compass bearings. The destroyer commanders concerned felt that Lieutenant Commander Wild's testimony had clarified their position clearly enough.

But now they were not so sure. The swift manner in which Admiral Pratt had placed Larry Blodgett in jeopardy was, to say the least, unsettling. If the Court were on a head-hunting expedition, their own heads might not be too firmly fixed. Captain Thomas T. Craven, who had been just one class behind that of Captain Watson at Annapolis, was counsel for the Squadron Commander. As Blodgett left the witness chair, Craven gave his client a bleak look accompanied by a shrug of the shoulders that spoke volumes without words.

The real contest over the validity of the radio bearing signals reached full proportions on the day Lieutenant Commander John M. Ashley, Superintendent of Communications of the Eleventh Naval District, took the stand. He was followed by G. C. Falls and F. H. Hamilton, radiomen on duty at the Point Arguello Station on the night of the disaster. Without skirmishing around the bush, Captain Craven delivered a direct thrust at the reliability of the Compass Station's bearing signals in the form of an accusing telegram.

It had been sent to Captain Watson by J. R. Stapler, who gave his address as 152 Riverside Drive, New York City. Captain Craven said that neither he nor Captain Watson knew the sender.

It was later ascertained, through the Merchant Marine Officer's Club, which is located at that address, that a J. R. Stapler had been First Mate of the freighter Arizona of the American-Hawaiian Steamship Line when this vessel left San Francisco on August 21.

The telegram read:

HAVE IN OUR POSSESSION TWO SETS OF RADIO BEARINGS WHICH, HAD WE NOT DISREGARDED THEM, WOULD HAVE PUT US ASHORE ON SAN MIGUEL ISLAND, AUG. 23.

ONE SET RECEIVED WHILE ANCHORED AMONG ROCKS ON NORTH SIDE OF ISLAND. WILL GLADLY FURNISH SAME WITH LOG EXTRACT, IF WILL BE OF SERVICE TO YOU.

Mr. Stapler's spectacular message gave the defense but a temporary lift. He had left the Merchant Marine Officer's Club for a visit in upstate New York. Thus Stapler could not be reached to back up his charges with sworn testimony before the Court.

When he was pressed for information concerning the conflicting 2035 radio bearing, Commander Ashley replied: "There is no record in the log of such a bearing!"

"Are you aware that three different persons copied such a bearing from the Point Arguello Station?" thundered Craven.

"And that Commander Hunter later saw a typewritten copy of it on a desk in the office of the Communications Supervisor of the Eleventh Naval District?"

Commander Ashley answered that he was not familiar with those allegations. Captain Craven next called attention to the report that one of the station's sending masts was down on the day of the disaster.

Ashley admitted the possibility of a very small error in bearings, due to the dropping of the mast. However, he said it would have no material effect on the general efficiency of the station's work of informing ships at sea in what direction from Point

Arguello they happened to be; that is, their compass bearings from that point. Much time was consumed in eliciting the exact number of minutes required to tell a ship its position after it had asked it of a radio compass station. The witness estimated it as 2 minutes.

The defense claimed that the time intervals on September 8 were much longer. One radio compass bearing may not always be the most reliable kind of a check on a ship's exact position, the witness admitted, but if three or four bearings are received, he said, the navigator should not hesitate to rely on their general accuracy.

G. C. Falls, Radioman 3rd class, admitted that he had sent a reciprocal bearing of 167° to the Delphy at 2:30 o'clock on the afternoon preceding the wreck—that is, he had wirelessed the vessel in effect that she was southeast of Arguello. When the bearing was questioned, however, he said he sent the direct bearing placing the ships to the northwest. Radio traffic between 6 o'clock that night and 9:05, when the Delphy and her sister-ships plunged into the rocks, was "abnormal," according to Radioman F. H. Hamilton, who testified he was the operator on duty at Point Arguello during those hours.

There was considerable interference in the air caused by the USS Seattle, he said, whose radio messages cut into the traffic so strongly that it was impossible at times to read messages from other ships. He flatly denied having received a message from the Delphy, saying: "We are to the southward; give us reciprocal."

Up to and past this very time of the hearings, strenuous efforts had been made by shore searchers and divers at Honda to discover the log of the Delphy, which was lost when the destroyer broke in two forward of the after deckhouse. It will be remembered that the forward portion of the ship became inaccessible immediately after Captain Hunter left the Delphy's bridge and went aft to execute the order to abandon ship. Thus he had no chance to save the ship's log and other valuable papers. Had the log been found, so the defense contended, it would have given proof of the sending and receiving of the disputed messages.

On the other hand, divers did recover the Delphy's navigation charts. During that first week of hearings, which began on Monday, September 17, the defendants and their counsel had met between sessions to discuss the gravity of the situation that had arisen as the result of Lieutenant Blodgett's being made a defendant because of his testimony regarding his part in the navigation of the Delphy. If this became a trend in the treatment of other witnesses, it would create a serious handicap to the defendants, because counsel for other defendants could not subject them to cross-examination.

After protracted discussion, all 13 defendants decided to waive their rights regarding the privilege of a defendant not to give testimony unless he so desired.

Watson Dons Mantle of Blame

When the Court met on Saturday morning, September 22, Captain Craven, as Senior Counsel, informed the Court of the rather momentous step taken by all the defendants. Said Captain Craven:

"Due to the fact that the Commanding Officers, Division Commanders, and Squadron Commander of the vessels that appeared in the incident under investigation are all defendants, it has made it difficult to bring before the Court navigational data and details to correctly picture the catastrophe. After the Navigator of the Delphy had testified to certain details in reply to questions of the Judge Advocate of the Court, he found himself placed in the status of a defendant and withdrew from the stand.

"Consequently, he became ineligible to cross-examination by the defense. It would appear that if this process should be continued and navigators of other vessels were called as witnesses, and, after replying to certain questions by the Court and the Judge Advocate, these officers became interested parties, the defendants would automatically be barred from presenting views of the disaster, other than those revealed by the interrogations by the Court and Judge Advocate.

"The complete picture would not be obtained, for the rights of the defendants would not be completely guarded. Captain Watson, from the beginning of this investigation, has been anxious to take the stand to explain motives which prompted certain decisions on the evening of September 8."

Speaking for the Court and the Prosecution, Admiral Pratt and Judge Advocate Bratton characterized the action as "worthy of the best traditions of the Navy and deserving the highest praise."

The Court then adjourned until Monday, September 24, to give the defendants time to prepare themselves to testify. Coming on a weekend, when the press contingent anticipated a let-down in the trial pyrotechnics, the Saturday morning session had not been given wide attendance. Newsmen who were at hand made the most of a statement by Captain Watson to the effect that he would assume full responsibility for the disaster when the Court met the following Monday. In a statement that was virtually a final draft of the announcement he would make before the Court, the Squadron Commodore said:

"The responsibility for the course of the Destroyer Squadron was mine, a responsibility which I fully realized. When I ordered a change of course, the change that resulted in the wrecking of seven destroyers of Squadron Eleven and the consequent loss of life, I did so fully cognizant of that responsibility. But that decision was based upon about three decades of experience in the Navy and made after due consideration of bearings obtained from the Point Arguello Radio Station.

"We had attempted, from time to time, to get bearings from the Point Arguello Station. But because of the extensive demands upon that station, probably due to the proximity of other ships in the vicinity, we were unable to get expeditious contacts.

"Due to overcasts, we were unable to get our true positions from the stars. When we received word from the Point Arguello Station that we were to the north of that station, I simply could not believe it.

"Remember we had traveled about 120 miles since we had received our positions earlier in the day!

"I asked for our bearing. From about 6:30 to 8 o'clock—that most vital period—we were unable to get radio bearings from the station. I had every reason to believe in my own mind that we were south of the Point. When we did get the bearings, they were confused. It seemed to me that they had made a mistake. I insisted that we were south of the Point and asked for a confirmation. The station insisted that we were to the north. There was only one thing to do and that was to make a decision as to whether or not the station was correct. Remember that there was much interference.

"I accept the responsibility. I was convinced that the station was wrong. But they were right."

In testimony given on September 24, following the submission of a formal statement which closely paralleled his newspaper interview, Captain Watson declared: (1) that he made the decision to order the 55-degree turn at 8:50 P.M.; (2) that he at that time instructed Lieutenant Commander Hunter, commanding the Delphy, to proceed upon the course until 9 o'clock; and (3) at that time to turn to the eastward on a course which assumed that the vessels would be safely south of Arguello Point and in a position to swing into Santa Barbara Channel. Captain Watson went on to explain: Shortly after 8 o'clock, a bearing had been received by the Delphy. It indicated that the vessel was some distance north of the Point. Since their dead reckoning showed the ship well to the southward of the Point, it was assumed that the radio operator had given a reverse bearing, so the much-disputed message requesting a "reciprocal" was sent at 8:25 by Hunter.

The answer came, after a delay of 10 minutes—at 8:35. The reciprocal was 168 degrees true. At that time, Captain Watson and Hunter were in conference about the course situation over the navigation chart in the Captain's emergency cabin abaft the bridge. The order to change course at 9 o'clock—or 2100—was decided upon in the light of the 8:35 reciprocal radio bearing which placed the Squadron south of Point Arguello.

For the sake of the record, Captain Watson recalled that Squadron 11 had been proceeding at 20 knots, standard speed for the cruise, with the wind moderate and the sea moderate to rough. Both wind and sea were astern. The speed of 20 knots had been ordered by Rear Admiral Kittelle as Commander Destroyer Squadrons.

Regarding the 8:35 radio bearing and assuming that the ship had been "set" inshore enough to place her on that bearing, she would at 8:35 have been south of Arguello, if she had made good at 20 knots, Captain Watson said. Thus, with dead reckoning and radio compass bearing apparently checking, all angles of the situation were discussed. At the close of the conference, the fatal order was given. There seemed to be no reason to expect an inshore current strong enough to be any menace to their course, as it had been laid to clear Point Arguello 9 miles west and well outside the 100-fathom curve. Answering a question from Admiral

Pratt, the Squadron Commodore replied that soundings could not have been taken while the Squadron was traveling at 20 knots; that a division commander would first have had to ask him for permission to slow down for soundings.

Again Admiral Pratt interrupted: "But, as a matter of fact, were your navigation facilities any better than those of your division commanders, had they chosen to exercise a little initiative?" he asked.

"No," answered Captain Watson.

"Do you feel that you can assume all of the responsibility that at times must fall on the shoulders of your division commanders, particularly that part of their responsibility which concerns the safeguarding of their vessels?"

"I have no desire to assume their responsibilities," came the slowly spoken reply.

"I simply want to make clear that I assume all of my own."

Suddenly, Admiral Pratt veered off on another tack that took the inquiry briefly, but pointedly, onto the high seas of Prohibition. In answer, Captain Watson denied emphatically that liquor was in any way responsible for the disaster. He added that, if any of his officers and men had had liquor aboard their ships, he would certainly have known about it.

"Destroyers," he said, "are such small vessels that officers and men are thrown into closest association, and one could not conceal liquor from the others."

In closing his testimony on that day, Commodore Watson reiterated his regrets at the loss of men, who, unlike ships, cannot be replaced, and the hope that the responsibility for the disaster (which he considered his own) would not descend upon the able and loyal subordinates who supported him on all occasions. Their actions in saving men and reducing the number of stranded vessels to a minimum, seemed to him to have exhibited initiative, skill, and decision of which any Navy might well be proud.

Captain Watson's assumption of responsibility had cleared the atmosphere. But it was evident from Admiral Pratt's questions on the subject of responsibility, that the Squadron Commander would not be allowed to carry the sole burden of guilt even if he were perfectly willing to offer himself for the sacrifice. On this score, correspondents wrote reams of dope stories which helped to keep the pot of public interest at the boiling point.

Lieutenant Commander Hunter testified that at no time between the departure from San Francisco to their arrival at a position off Pigeon Point was any ship in the Squadron steering a steady course or making a steady speed.

Consequently it was impossible, he said, to check with accuracy to revolutions per minute of their propellers, although they based their estimate of distance covered on this revolutions count. During his turn on the stand, Hunter was subjected to extensive cross-examination after he had given an explicit account of the, to him, confusing radio bearings. He had just finished citing an unseasonable northerly current and the type of radio compass used at Point Arguello as the real causes of the destroyer disaster when he was interrupted by Admiral Pratt, who asked acidly: "Do you mean to say that the wreck was an act of Providence or an error in judgment?"

"I'll have to admit that it was an error in judgment," Hunter replied.

He was somewhat taken aback by the question, but continued:

"Still, as contributory causes, I believe the unusual northerly current we encountered near Point Arguello and the fact that a bilateral radio compass is used there were partly responsible. I think there is also a possibility that abnormal currents caused by the Japanese earthquake might have been another contributory cause, or magnetic disturbances connected with the solar eclipse affecting the compass; but of these I cannot, of course, speak with any first-hand knowledge."

The mythology of wayward currents was ridiculed by Captain N. P. Cousins, master of the Ruth Alexander, a famous coastal passenger vessel. He said that, based on 30 years of experience in traveling along the California coast, he found it safe to pass within 2 miles off Point Arguello. As for currents:

"In some months," he said, "the current is with your ship going north; in other months, it is with you going south. You simply have to know the right time of year."

With respect to negotiating the region of Honda in foggy weather, he sagely observed that the best way of checking his course was to heave the lead and take frequent soundings. But when it came to placing reliability on radio bearings, the veteran skipper was solidly in Hunter's corner. "We usually take bearings every half-hour or so in a fog. But we do not rely upon them. They are wrong as often as they are right."

With respect to Captain Hunter's belief that magnetic disturbances might have been contributory causes, it should be mentioned here that there is no problem of magnetic variation or deviation at a radio compass station.

The compass circle is adjusted to true north and the bearings given to the ships are arrived at by the ship's sending a series of long dashes while the compass operator ashore centers his receiving apparatus on the sound and reads off the true bearing of the ship from the station. Of course, the perfection of such a system depends on the binaural sense of the operators, but these have been tested by Navy medicos before assignment to such jobs. Nearby electrical circuits or buildings might affect bearings and, of course, every compass has to be carefully calibrated. It took the press and public a day or two before they caught on to the full scope of the magnanimity of Captain Watson's willingness to shoulder the responsibility for the pile-up.

An editorial in the Army Air Navy Journal gives a thorough analysis of this "generous action."

The editorial follows:

Whatever may be said as to the degree of culpability of Captain Edward H. Watson in connection with the recent destroyer wrecks on the Pacific coast, nothing but praise can be given him for his outstanding manliness in assuming full responsibility and in seeking to relieve from blame those whom he terms his able and loyal subordinates. This is a type of genuine leadership of which the United States Navy may well be proud and in which it will find much strength. Captain Watson has not merely to consider the possible immediate punishment which may come to him as a direct consequence of the affair.

Under the system of promotion by selection in the Navy he is also confronted with the probability of a permanently ruined career in the event of his serious culpability being established.

Under such circumstances he could never hope for further promotion; never hope to achieve that desire strongly inbred in every naval officer; to serve the country well in a high capacity.

Yet Captain Watson has given a splendid example of the finest attributes of character overcoming the elemental instinct of self-preservation.

Voluntarily waiving the fundamental right of a defendant to place the burden of proof upon the prosecution, and to refrain from testifying under oath to any facts which might tend to incriminate himself, he took the witness stand and not only freely testified to facts relating to his own culpability but also volunteered his opinion under oath that he was wholly responsible for the disaster, and that none of his subordinates should be blamed.

This is indeed worthy of the best traditions of the Navy, as Rear Admiral Pratt is quoted as having said. Leadership inevitably involves responsibility. It is easy to accept such responsibility for good results.

It requires a full measure of character to voluntarily accept it for bad results. Examples of the latter are inspiring and cannot fail to build up service morale, that quality of teamwork and of spirit which transcends in importance every other factor in the highly difficult and complex part of war efficiency. The example of Captain Watson in this regard is a personal contribution to such efficiency, difficult to exceed in time of peace.

Whether or not he may be present at some future fleet battle, regardless of the outcome of the present legal procedure, he will have contributed much towards victory. In San Diego, Captain Robert Morris, ComDesDiv 33—which included the S. P. Lee, Young, Nicholas, and Woodward, all consigned to the bottom of the sea—testified that he had taken all possible precautions for the safety of his division on the night of the disaster.

He also maintained that commanding officers of destroyers that steamed in column astern of the Squadron Flagship "could not possibly be held culpable in carrying out the destroyer doctrine of following their leader."

"We have frequently had to follow our squadron leader where no course whatever was signaled and we were not given any information about the course," Captain Morris testified.

It would have been "very much out of place" for any officer in the Squadron to have suggested to Captain Watson that the Squadron's speed be cut down for the purpose of taking soundings, said the witness, inasmuch as he was their superior officer, years ahead of them in seniority.

"Does seniority ever take the place of common sense?" flashed Admiral Pratt in a voice that had the edge of a naked blade.

"They are supposed to be synonymous, sir," replied Captain Morris placidly.

Under further questioning by Admiral Pratt, Captain Morris told how his division ran into a thick fog in Puget Sound that summer and was saved from disaster only by a quick reduction of speed and immediate resort to soundings.

"If you had been in command of Destroyer Squadron Eleven on the night of September 8, would it have gone on the rocks?" asked Admiral Pratt bluntly.

"I do not like to flatter myself," responded the witness, "but inasmuch as I am testifying under oath, I will say no. It would not have gone on the rocks."

By and large, the hearings proceeded on a level of necessarily repetitious technicalities, and the news material was rather staid despite the heroic efforts of the sensation seekers to whoop it up.

They had a field day, however, when Commander William Calhoun, Captain of the Young, took the stand, and raw emotions swept technicalities out on a tide of tears.

Bill Calhoun, a very devout man, who had taken the loss of his men intensely to heart, broke down as he read the list of 20 of his crew who perished when the Young turned over.

His testimony, like that of most other witnesses, began in a mass of technicalities involving courses steered, bilateral radio compasses, the approximate set of an inshore current, and the like; but soon his words painted pictures of men in peril of sudden death, of ships breaking up in a tumult of rocks and surf, of discipline prevailing over fear, and of heroism rising above the instinct of self-preservation. He told how the Young, slashed wide open by the reef into which she had crashed at 20 knots, capsized within a few moments of striking, and how, despite this sudden and complete disaster, all but 20 of her crew were saved. All of his men were heroes that night, he said.

To a man they preserved that discipline without which practically all of them must have been lost in the hell of surf and steel into which a fatal change of course had thrown them. He commended Captain Watson.

"I only hope that if ever I am faced with the tragedy that faced him that night," he said, "I'll be half the man that he was: cool, calm, courageous, and thoughtful; never missing an opportunity to aid."

As he began to read the list of men lost in the wreck of his ship, as is customary at inquiries of this sort, Calhoun's voice choked.

He stopped frequently. Tears rolled down his cheeks.

When he had finished reading, Captain Calhoun turned to Admiral Pratt. "Admiral," he began, "I want to apologize for letting my feelings get away from me—I—."

Then he broke down again and fellow defendants came forward and led him away.

Bearings "Inaccurate but Correct"

By October 1, all the defendants had completed their unhappy pilgrimages to the witness chair. They were followed on that day by a strong proponent of radio compass bearings in the person of Lieutenant Commander H. A. Jones. His testimony, in view of his standing as a former instructor in navigation at Annapolis and a salt-water sailor who had logged 7 years as a commander of destroyers in peace and war, was effective.

According to Commander Jones, all of the radio bearings received by the Delphy shortly before she led Destroyer Squadron 11 into the rocks off Honda, were inaccurate. But, he added, they were all correct in so far as they showed the ships still too far north to attempt an eastward change of course in the Santa Barbara Channel. The bearings were from 1 to 8 degrees in error, according to the witness. Fifty miles out to sea, he explained, such an error would destroy the usefulness of a radio bearing, but at the distance from land at which the Delphy was traveling, they were "accurate enough to indicate her approximate position."

In the approximately 25 instances in which he had taken a vessel past Point Arguello en route to San Diego, the witness said, he had not hesitated to use radio bearings, despite the fact that "they have generally been in error," a few degrees one way or the other.

So reliable in a general way were these bearings, he said, that he frequently depended upon them when rounding the Point in foggy weather and did not think it necessary to take soundings as an additional check on his navigation. Had he been in command of the Delphy on the night in question, however, he added, he would have begun taking soundings at 8 o'clock, an hour before the fatal change of course was executed, in an effort to determine the squadron's exact position.

A side comment seems to be in order at this point. Squadron 11 and Squadron 12 were on a competitive 20-knot engineering run. To slow down for soundings, the Delphy would have had to reduce her speed to about 5 knots. All the other vessels in the column would have had to make similar slowdowns to maintain station. Under the spur of competitive pressure—plus the assurance they felt that their dead reckoning was right and the station bearings wrong—it is not too surprising that the Delphy did not slow down for soundings. One might say it was not exactly the destroyer way to resort to frequent castings of the lead. Unlike Lawyer Marks of Uncle Tom's Cabin fame, the style of destroyermen was not to make sure they were right before they went ahead. They knew they were right. And they went ahead.

And that cocky air of self-assurance has always been the true destroyer spirit as well as the foundation of its efficiency. There were, of course, exceptions. Even in Squadron 11 on that night of September 8. Such a one was Lieutenant Commander Thomas A. "Tommy" Symington, captain of the Thompson, last destroyer in column. In his testimony, Tommy Symington demonstrated the value of using the lead line, and also of using his own judgment, when he found his ship in jeopardy.

He testified that he saw ships ahead of him turning east about 9 P.M., 10 minutes before his dead reckoning showed he would be abreast of Point Arguello. When the change of course was followed by a confusion of lights and whistling of sirens, he decided not to make the turn and instead slowed down to take soundings. These soundings, he said, showed the Thompson was in 10 fathoms of water, so he turned out to sea instead.

Congress Blamed for Wrecks Captain Tomb, Commodore of Destroyer Squadron 12, against which DesRon 11 competed during the 20-knot engineering run that had such a disastrous ending, was a spectacular witness.

He put his cards openly on the table when he declared that, if Congress had not refused to grant the funds for proper equipment of destroyer squadrons, the disaster would never have taken place. The Navy Department had tried repeatedly, but without avail, to obtain appropriations for the construction of ships properly equipped to act as destroyer squadron leaders, Captain Tomb asserted. Instead, ordinary destroyers, with insufficient quarters for the Squadron Commander's staff and carrying no sonic depth-finders with which to take soundings, were still used for this purpose.

"Had a squadron leader of the type used in the British navy been assigned to Squadron 11 on September 8, no such disaster would have occurred," declared the Captain, who had been called to the stand to tell how his Squadron had rounded Point Arguello in safety, while seven ships of the Eleventh steamed off their course and into the rocks. Under cross-examination by Rear Admiral Pratt, the witness admitted that—had he been in command of Squadron 11 and in possession of the data available to its navigators—it would not have gone on the rocks.

Even without a squadron leader of the British type. He reiterated, however, his previous assertion that with a vessel of the British type at the head of the column, equipped with a sonic depth-finder for soundings, the Eleventh Squadron would not have met disaster.

The Radio Compass Station at Point Arguello, whose direction signals were disregarded by navigators of the Delphy, also came in for criticism by Captain Tomb.

"It is amazing," he said, "that a compass station serving such a dangerous stretch of coast line is not equipped to give ships at sea a prompt fix whenever requested, as can be expected from similar stations on the East Coast. With the radio operator at Point Arguello harassed and hurried by a multitude of requests for bearings, as he was on the night of the wreck, it is too much to expect that the resultant bearings should be accurate."

Bearings received from Point Arguello by his Squadron that night, said the witness, were from 3 to 10 degrees in error.

"As a matter of fact," he added, "the radio bearings gave us practically no information on which to determine our exact position."

"But as a matter of practical navigation," asked Admiral Pratt, "admitting that the Point Arguello bearings were ten degrees in error, did or did not these bearings assist you in keeping off the rocks that night?"

"Yes, they were a very vital factor in causing us to change our course," responded the Captain.

"Isn't that the very purpose of radio compass bearings?"

"Yes, it is," admitted Captain Tomb a bit sheepishly.

The compass bearing—which led him to change course into safer waters at almost the same instant that Squadron 11 had changed course into the rocks—was "a lucky bearing," the witness admitted, in that it was only 1 degree in error. Had he not changed course, at that time, he said, his Squadron might have run into trouble. It was a "very abnormal current" that placed his ships in such a critical position, he said, explaining that they were more than 8 miles from the point where their dead reckoning showed them to be.

Previous witnesses testified that the Delphy was a little over 11 miles off her course when she led Squadron 11 into the fatal eastward turn at 9 o'clock.

Reporters made a lot of news-hay while this sun of contradictory testimony of "yes—buts" was shining. But a last-day witness, Commander Walter G. Roper, ComDesDiv 32, filled their hayloft to overflowing. He was called as a witness on October 11. This, appropriately enough, was the last day of the Court of Inquiry, and doughty "Uncle Walt" was one of the last on the stand. When asked by Admiral Pratt if he had intended to follow the Delphy around a change of course at 9 o'clock on the night of September 8, Commander Roper replied tartly: "I wouldn't have followed the leader around a sixty-five-degree turn. I don't believe in following the leader if the leading boy jumps off the barn."

"If you had observed the Delphy turn to the left, would you have tried to stop her?" queried Admiral Pratt with beguiling mildness.

"Yes, if I see a man trying to jump out of a fifteen-story window I'd stop him if I could. You can't turn a corner until you reach it, and the Delphy had not reached the corner."

"And how did you arrive at that conclusion?"

Roper explained the movements of his ships during the night of September 8 and on the following morning. None of the destroyers in his division, the last in the column, was damaged.

"It occurred to me that the ships would not make the speed by engine revolution," he continued. "The ships were yawing badly and that would tend to hold them back. I've noticed that destroyers never make the speed indicated by the turns on the propeller—never. I think it bad practice to get the speed that way. The patent log is the way to get the speed."

"What speed did your patent log show the Kennedy [Roper's flag-ship] was making at eight o'clock that night?" asked Captain Sellers.

"It happened we didn't have the log out that night," Commander Roper replied. "Patent logs must be calibrated," he explained, "and our log hadn't been calibrated correctly."

Radio bearings taken by destroyer position finders are fairly accurate, Commander Roper declared in response to a question. He was shown a publication in which some Navy officer had declared radio bearings taken from ships are only rough approximations.

"I don't know who wrote that," Commander Roper said.

"Maybe it was written by someone who hasn't taken as many radio bearings as I have. I'm not much of a desk man. Most of my knowledge has come by hard experience."

Commander Roper thought his ship had experienced an inshore set, but seemed skeptical when questioned regarding currents.

"Many so-called currents are due to mistakes in dead reckoning caused by bad steering and other such unfavorable conditions," he observed. Amazingly enough, when Roper found, by radio bearings, that his ship was proceeding considerably closer to the shore than expected, he felt he had no cause to issue a warning to ComDesRon.

"It never occurred to me that other ships did not have the same information I did," he said. "I expected we were running for Arguello Light to get a definite fix on our position. I thought that was beautiful, superb navigation. You couldn't beat it."

Roper declared that he wanted to get a fix on Arguello Light himself and then go out in search of survivors reported adrift in a boat from the wrecked steamer Cuba on San Miguel Island. For that reason he did not think the speed excessive. "If I had a hundred knots instead of twenty, I would have used them to go out to the Cuba as soon as I saw the Light," he testified.

A radio telephone conversation between himself and Captain Watson was described by Commander Roper. He said he called up ComDesRon to suggest a plan for searching for a boat said to contain survivors of the Cuba. Captain Watson replied that he was unprepared to undertake such a major operation in view of the fact that the Destroyer Force Commander was in the vicinity.

When red lights showed on destroyers ahead and the column appeared to slow down, Commander Roper said, he already had arrived at a decision to sheer out to the westward. He thought at the time that there had been a man overboard as he had not noticed the change of course.

"Did you notice the confusion ahead?" the Judge Advocate asked.

"No, I didn't notice confusion. I thought it was very well done," Roper replied.

"I saw a flare on the water and I expected the ships were stopping and backing to pick up the man overboard."

Following brief arguments by the Judge Advocate and counsel for the defendants, the hearings came to a close. The Court of Inquiry adjourned to consider the evidence and arrive at a decision that would affect each of the 13 defendants who had appeared before it. The conduct of the inquiry had been fair but tough. The 13 men, whose careers might be destroyed by the tilt of the Scales of Justice, were sure of one thing—those who sat in judgment had not been blind.

REAR ADMIRAL SUMNER KITTELLE (1867-1950)

12 - Doomsday in San Diego

In the Opinion of the Court

ON OCTOBER 31 A VERITABLE BLOCKBUSTER—that shattered the serenity of the nation's peacetime Navy—was dropped from the august heights of the Navy Department in Washington by Secretary Denby. It was the publication and endorsement of the findings and observations reached by the Court of Inquiry.

Eleven officers were recommended for trial by General Court Martial. Captain Edward H. Watson, ComDesRon 11; Lieutenant Commander Donald T. Hunter, Captain of the Delphy, Flagship of the Squadron, and Lieutenant (jg) Lawrence F. Blodgett, Navigator of the Delphy, faced recommendations that charges of culpable inefficiency in the performance of duty, and through negligence suffering vessels of the Navy to be run on the rocks, be brought against them.

While the foregoing findings were not wholly unexpected, the recommendations—that two Division Commanders and six destroyer Captains stand trial on charges of negligence, in suffering vessels of the Navy to be run on the rocks—exercised a pronounced shock effect. What had happened to the follow-the-leader concept of the long-established but unwritten destroyer doctrine? Would this, officers asked themselves and each other, bring about drastic changes in well-tried theories of strategy and tactics?

Those recommended for trial on these grounds were:

Captain Robert Morris, ComDesDiv 33, and Commander William S. Pye, ComDesDiv 31; Commander Louis P. Davis, Captain of the Woodbury; Commander William L. Calhoun, of the Young; Commander William S. Toaz, of the S. P. Lee; Lieutenant Commander Walter D. Seed, of the Fuller; Lieutenant Commander Herbert O. Roesch, of the Nicholas and Lieutenant Commander Richard H. Booth, of the Chauncey.

Eliminated from the action were Commander William P. Gaddis, of the Somers, and Lieutenant Commander John F. McClain, of the Farragut. Their vessels were but slightly damaged as they extricated themselves from the Devil's Jaw.

The opinion, which reviews all the facts and circumstances in interesting detail, follows:

In the opinion of the Court, the disaster which resulted in the stranding of seven destroyers on Pedernales Point, and the grounding of two others in the same vicinity, is, in the first instance, directly attributable to bad errors of judgment and faulty navigation on the part of three officers attached to and serving on the Delphy, viz., the Squadron Commander, Captain Edward H. Watson; the Commanding Officer, Lieutenant Commander Donald T. Hunter, and the Navigating Officer, Lieutenant (jg) Lawrence F. Blodgett.

Based on the testimony adduced, and to determine the degree of responsibility to be borne by the Squadron Commander, the Captain of the Delphy, the Navigating Officer of the Delphy, the Division Commanders of the 33d and 31st Divisions, the Commanding Officers of the seven stranded ships, and the Commanding Officers of the two ships which touched the ground, but which were not seriously damaged, and also to determine upon other matters connected with the court's investigation, the Court has found it fit to divide the period from the time of passing Pigeon Point abeam to the time when the last of the crews of the stranded ships were landed, into three periods, viz.:

A—The period from the passing of Pigeon Point abeam to the time of the turn of the head of the Squadron to the left on a course of zero-nine-five, which took place at about 9 P.M., Sept. 8, 1923; B—The period from the time of the turn of the head of squadron column, at about 9 P.M., to a course of 95, to the time of the stranding of the last ship, viz., the Chauncey, in all a period of approximately 6 minutes; C—The period which elapsed from the time of the stranding of the last ship, the Chauncey, to the time on the following day, Sept. 9, when the crews of the Fuller and Nicholas, the last ships to land their crews on shore, were able to so land them.

In the opinion of the Court, safe courses of 160, from the fix off Pigeon Point, to pass Point Sur, and 150 to pass Point Arguello were safe courses to set. The speed 20 knots during the period "A" was not an excessive speed. The D.R. course and the speed as shown by revolutions were not made good, but the

Squadron was during this time set in to the coast, and north of the D.R. position at 9 P.M. by a very appreciable amount.

The Court is of the opinion that no unusual current conditions existed, but that this set to the north and east was caused by bad steering, together with a certain amount of current which, while not explicitly laid down in the Sailing Directions, may be expected at any time in any direction and should be guarded against by the careful navigator.

During this period certain radio compass bearings were taken by the Delphy, but the fact that these bearings were not transmitted to the ships following constitutes a neglect on the part of the Squadron Commander, who should have seen that adequate information, to insure the safe navigation, was transmitted to them.

However, upon not receiving information to safeguard the navigation of their own divisions, it then became the duty of the Division Commanders to ask for such information from their Squadron Commander, or to take such independent action on their own part as would insure the safe navigation of their own units. Particularly did this duty devolve upon the Second in Command who should at all times possess himself of the required information to take over the command of the entire unit and to conduct it to safety.

To a lesser degree this same comment applies to the next in command and to the captains of each individual destroyer.

Period "A" was the critical period, for upon the information possessed by each captain and by each Division Commander depended his ability to make a correct and accurate judgment of what his action should be as he neared Point Arguello, which was the turning point of the Squadron. The Court believes that too much stress was laid by all ships upon the 6:30 P.M. radio bearing, because it checked with the D.R. positions. The critical period following between 8 and 9 P.M., when great stress should have been laid upon the receipt of compass bearings, was neglected by all ships following the Delphy, because (1) they placed too much confidence in the Delphy's 8 P.M. position, (2) because they put too much faith in the 6:30 P.M. intercepted compass bearing.

This opinion holds for the ships of the 33d and 31st Divisions following the Delphy. The opinion of the Court is that the

32d Division was at all times in possession of navigational information sufficiently complete to enable it to operate safely under all conditions. The Delphy presents a curious case.

Confident in their own D.R. and discrediting the compass bearings because they were thought confusing, the attitude of mind of the Squadron Commander, the Captain of the Delphy, and the Navigating Officer of the Delphy, was one of complete assurance, at the very time when a doubtful situation had arisen.

This situation arose between 8 and 9 P.M. and became particularly acute when the Delphy sent the signal, "We are south of Arguello," and asked for a reciprocal bearing. The safe procedure at this time would have been to reduce speed, take soundings, and proceed cautiously until further radio bearings had approximately fixed the position of the leader. This was not done, and it indicated a state of confidence of mind of the leader which was naturally imparted to those following, who had not through their own individual efforts fixed the position of their own ships. The result of the above procedure was that the Delphy, the 33d and the 31st Divisions arrived at the turning point at 9 P.M., the Delphy with information inaccurately interpreted and the 33d and 31st Divisions with insufficient information to enable them to quickly take appropriate immediate action when such action was imperative.

Had the responsible parties on the Delphy not assumed her position, through bad errors in judgment and misplaced confidence, to be South of Arguello, but had proceeded further on the course 150, it is the opinion of the Court that Arguello Light or the fog signal would have been picked up ahead or very slightly on the port bow, a position which was not unsafe, as the Squadron could have been maneuvered quickly to the clear water westward.

Particularly would this be true had the speed of the Squadron been reduced at 8:30 P.M. and soundings taken. The necessity of obtaining a fix at Arguello was apparent as the Squadron was to proceed through Santa Barbara Channel, where fog might be expected at any time.

This would have been good navigational procedure. Instead of this procedure, the Delphy without a proper fix turned sharply and blindly to the left to the course zero-nine-five, at a speed 20, and at a distance from the shore as plotted on the official chart of less than 1½ miles from the reefs off Pedernales Point. In sequence the 33d and 31st Divisions arrived at the turning point and made the turn with the Delphy. At this time, or within the next minute or two, and only at this time could disaster have been avoided by the ships following the Delphy. It is doubtful whether the Delphy could have been saved by any action on the part of any following ships, for the Delphy up to this moment had not indicated her intention to turn, her speed was too high.

The officers responsible for her safety were too obsessed with the idea that they were south of Arguello Light, and the distance to the rocks was too short. A signal could not have gone through in time, in all probability. But it is believed that had, at this time, the 33d and 31st Divisions stood on, instead of turning to the left, or had they turned sharply to the right, the 33d and 31st Divisions would have been saved.

This action on their part would not only have been good, but it was imperative, and it was necessary that it be made immediately and without delay. It is probable that this action would have been taken had those two divisions arrived at the turning point with all the information at hand then in possession of the Delphy, obtained by intercepting bearings, or if this method proved inadequate by directly asking the Delphy for this information, or failing in that by taking bearings themselves as done in the 32d Division, even though it was against rules. In the opinion of the Court no rules or regulations, no formal practice of guarding set radio waves, may preclude a Captain or Division Commander from taking every navigational precaution to safeguard his own ship or division, as was done by ships in the 32d Division and in the 12th Squadron.

He must risk a rebuke instead, and must at all times be prepared to take the initiative and to use his own individual judgment. The Squadron Commander erred badly, very badly in judgment, but his errors must not be allowed to creep down the line even at the risk of severe rebuke from the senior in command.

The senior may even welcome a suggestion which gives him a point of view other than his own, particularly when he is in a doubtful or hazardous position himself. Having straightened out on the course zero-nine-five, at speed 20, the position of the Delphy and of the 33rd and 31st Divisions was hopeless, the danger being greatest to those ships nearest the head of the column, and in proportion as they were near the head.

Led by the Delphy straight for the bluff on Pedernales Point with the coast, and reefs on the left hand, with the bluff ahead, and with an outlying island and sunken reefs on the right hand, neither a turn to the right nor to the left could save these ships.

The 33rd Division, like the Delphy, was doomed and crashed on the rocks and reefs to the right and left. Only by the promptest of action in backing, induced by the catastrophe to the ships ahead, was the Division Commander of the 31st Division able to minimize the disaster to his Division and reduce the total losses in his Division to two ships. Even this action on his part could not save the Fuller which ship, less lucky than the others, had already struck a sunken reef at the moment she tried to back.

Quick and prompt action on the part of the Captains of the Farragut, Somers and Percival alone saved their ships, though the Farragut and Somers touched. The damage, however, was not sufficiently great to render them unseaworthy or so great in extent as to render them unserviceable for any great length of time. The Chauncey stood on and was stranded due to the Captain's not knowing that the accident was stranding and not collision. Had the Squadron Commander sent a signal "I am aground," instead of "Nine turn" and "Keep clear to the westward," which conveyed inadequate information, it is possible that the Chauncey might have been saved.

At least the situation would have been clarified to the entire Squadron. The Commander of Division 32, which Division was in the rear of the column and therefore most favorably situated to avoid the disaster, was also by reason of more complete information better able to cope with the situation which developed immediately following the turn of the Squadron head at 9 P.M. By good judgment, good common sense, good navigational procedure, and by the good luck of being at the end of the column, the 32d Division turned individually to the westward to safety.

It is further the opinion of the Court that with the knowledge in the possession of the Division Commander of the 32d Division, and of the other ships in this Division which was this, viz.—that the Squadron had been very much set inshore and to the northward—that had this Division arrived at the turning point at 9 P.M. it would not have followed the Delphy in to sure destruction, and the present disaster would have been minimized. For his common sense and alertness the commander of the 32d Division is to be commended.

Periods "A" and "B" reflect no credit upon the Navy.

They were the periods when:

(1) A grave error of judgment was committed by the Squadron Commander, an error which practically caused the stranding of seven ships.

(2) When a too blind faith on the part of the ships following was placed in the judgment of the Squadron Commander.

(3) When too little initiative on the part of the ships following in the matter of determining their own independent positions was displayed. The disaster woke Squadron 11 up.

From that instant on, the Destroyer Force, Squadron 11, displayed a zeal, courage and coolness in face of grave danger, which is a matter of pride to the Navy, and should be to every American.

From the Squadron Commander down to the humblest man on board there was perfect discipline and the highest traditions of the service were lived up to.

Not a single case on any ship occurred where officers or men faltered in their duty, or failed to act calmly or coolly under orders. It is due to this perfect discipline that the loss of life in this disaster was so small.

The loss of life was confined to 23 men—20 from the Young, which turned over in about 1½ minutes, and 3 from the Delphy. They were lost probably at the time of the crash or shortly afterward. The crew of the Fuller, the most exposed ship, was landed the next day under most trying circumstances, not a man being lost. The crew of the Nicholas remained on board and were landed the following day.

The crews of the Chauncey, Woodbury and S. P. Lee were landed quietly and calmly when it became necessary to abandon ship to save life. Not a single attempt was made to abandon ship until the ship itself was a helpless wreck, and then only when it became necessary to do so in order to save life.

Every precaution was taken and the work was done quietly, orderly and efficiently. Had this not been the case, it is probable that the loss of life in this disaster would have been much greater due to confusion and possible explosion of the boilers on some of the stranded ships. This part of the story reflects the highest honor upon the Navy. It is a story of coolness, calmness, bravery and discipline in the face of grave danger.

The conduct and bearing of the Squadron Commander and that of all officers connected with the disaster was at the time and has been since during the entire conduct of the Court of Inquiry beyond reproach and deserving of the highest praise. Recommendations follow for officers and men deserving special mention for conspicuous actions.

After considering carefully the testimony adduced, the Court finds nothing which reflects upon the efficiency of the radio compass installation. A mass of confusing testimony has been brought forward to prove that bearings may not be relied upon, but out of this testimony shines the clear fact that it was not the compass bearings sent to the Delphy which were wrong, but the judgment of the men who interpreted these bearings and who used them wrongly.

All the night of Sept. 8 the 32d Division cruised to the north and south of Arguello Light in a heavy fog, working their way through the agency of radio compass bearings and the use of the lead. Up to 8:30 P.M. the speed of 20 knots on the course of 150 true was not to be considered excessive.

The weather conditions were not such as to make this speed dangerous from the point of view of seamanship and danger of collision with other ships. Destroyers handle better at this speed, and current has less time to act on them and drive them from their set course.

After 8:30 P.M. speed should have been reduced in order to take soundings, and after 9 P.M., if it had been necessary to head into the land in order to get a fix off Arguello, speed should have been reduced to a minimum until the Light was sighted or heard, or an approximate fix obtained through the agency of radio compass bearings, checked with soundings.

Only so long as the course was parallel with the land and not too close in, or away from the land, was the speed of 20 knots a safe speed, but it was safe so long as it did parallel with the land, or was away from the land until the fog shut in, when it should be reduced. An attempt was made to show that the principle of follow the leader was so fundamentally a part of the destroyer doctrine that to depart from the practice was always a grave error on the part of unit leaders unless they had information in their possession which warranted their so doing. This is no doubt true when the leader is right.

A departure from policy, plan, or even a strategic conception is rarely permissible, but in the tactical execution of the above much latitude must be allowed the subordinate.

Not only must it be allowed but the subordinate must take this initiative on his own responsibility when his judgment tells him this is the correct course to follow. The matter of navigational procedure comes more nearly under the head of tactical procedure. The Division Commanders and individual ship Captains are always charged with the safety of the unit under their command no matter who leads, unless it be in the presence of the enemy; when destruction of the enemy and not the safety of your own unit are the guiding factors.

No destroyer doctrine ever advocated the blind following of any leadership. On the contrary, the primary and strongest fundamentals are: loyalty to the plan, and in this way loyalty to the leader if the plan be correct; exercise of sound judgment on the part of the subordinate in carrying out the plan; development of the initiative on the part of the subordinate in order that the plan may be most efficiently carried out.

Had Nelson at Cape St. Vincent blindly followed the leader, John Jervis would not have gained the victory which he did. Had Nelson obeyed Parker, Copenhagen would not have been the monument to the British navy that it is. Blindly following the leader or unreasoning adherence to set regulation is more in accordance with the practice of those leaders of the past who hesitated to depart from the line ahead, even when advantages would accrue from a departure from such practice.

The plan on Sept. 8 was to proceed to San Diego. The procedure at the time of the disaster was a movement in column formation. It mattered little so far as detail went how the plan was executed so long as it was effectively carried out and the tactical details of execution were not at variance with the policy of the leader or would cause him embarrassment.

Nothing can replace the use of sound common sense on the part of the subordinate, and if he is not furnished with sufficient information by his leader to absolutely safeguard his own unit, or to effectively carry out the plan, he must ask for it himself, and failing in this, he must use every effort of his own to obtain it in order to better execute the general plan and by so doing aid the efforts of the leader.

This is imperative and is believed to be much more in accord with destroyer and fleet doctrine than to blindly follow the leader. Through unknown or unforeseen circumstances the leader may frequently err, as was the case when Tryon gave an order which resulted in the collision of the Camper down and Victoria, with the loss of the latter with great consequent loss of life.

In reviewing the testimony, the Court is forced to the conclusion that no unusual conditions existed. It is true that there was fog and that the lights which served as navigational aids were difficult to make. It would have been better practice to have made Point Sur and thus to have obtained a later fix before approaching Point Arguello, which was to be turned at night. It is true that a strong northerly wind would have helped to augment the speed of the Squadron, but this might be offset by bad steering. It is true that with the wind as it was a southerly set might reasonably be expected, but the sound navigator never trusts entirely to the obvious. The price of good navigation is constant vigilance. The unusual is always to be guarded against, and when the expected has not eventualized, a doubtful situation always arises which must be guarded against by every precaution known to navigators, such as the use and correct interpretation of radio compass bearings, and particularly by the use of the lead and a proper reduction in speed. When you cannot see, you hear and feel, until you are sure. The currents on this coast are so variable and so unreasonable in their actions that they cannot be relied upon definitely, and no ship is safe when close to the coast unless it actually knows where it is.

Dead reckoning alone can never be relied upon. It is always the captain who is sure in his own mind without the tangible evidences of safety in his possession, who loses his ship. In the opinion of the Court there is nothing which will excuse the Squadron Commander—the Captain of the Delphy—and the Navigating Officer from accepting the full responsibility for the accident.

Their responsibility is full and complete and the Court sees no extenuating circumstances. In the case of the Division Commanders the Court finds that they must be held responsible in a measure. It is true that they were following in column, that they could not slow, that they could not sound, that they could not without permission ask for compass bearings. But the fact remains that they did too blindly follow the judgment of the Squadron Commander, that they did place too much reliance upon the 8 P.M. position of the Delphy, that they did place too much reliance upon the one bearing about 6:30 P.M. which unfortunately checked too closely with the dead reckoning position.

Between the hours of 8 and 9, the critical hours, they did not check on their own initiative the actual position of their own units, and the result was catastrophe. It was possible to ask the Squadron Commander for information.

It was possible to unguard the battleship wave and the destroyer squadron wave and thus intercept the bearings sent the Delphy. It was possible to ask to take bearings themselves. It was possible to unguard the battleship and the squadron wave and use their own direction finders. The amount of traffic from Arguello, between 8 and 9 P.M., was sufficient for them to have obtained a rough bearing without interference to their leader and thus to have obtained information which would serve to give them an approximately correct estimate of the situation when the Delphy turned shoreward at nine, and that might have enabled them to make a correct and accurate decision of their own at that time. The responsibility of the Division Commanders is much less than that of the Squadron Commander, the Captain and Navigator of the Delphy—but it is a responsibility nevertheless. Due to the subordinate position they held it was a responsibility which did not involve malfeasance, but it does involve nonfeasance to a limited degree.

Concerning the individual captains of the stranded ships, their case is a peculiar one. They were not only following the Squadron leader, but they were also following their Division Commander and in the case of the S. P. Lee and Farragut, the Division Commander was personally on board these ships. Their position was a trying one and there are many extenuating circumstances. They had two leaders whom they must go through before they could slow, take soundings, change course or even ask for radio bearings. They did attempt to intercept radio compass bearings as was shown in the testimony, but unfortunately the two messages upon which most faith was laid were the two most misleading ones, viz.: the Delphy's 8 P.M. position and the 6:30 P.M. compass bearing. They did not know that the Delphy's position was one by dead reckoning and not by fix.

However, to a lesser degree, the same responsibility, which rests upon the shoulders of the Division Commanders, rests upon them. In the Period "A" up to the 9 o'clock turn they were not in the possession of information necessary for them to take action radically out of the ordinary, which action was imperative at 9 P.M. or shortly after, if they were to save their ship. In other words, it was necessary for them to take steps on their own initiative to obtain the information which they should and probably would have obtained had they been acting singly, but which they did not deem so essential in the presence of their seniors.

In this they erred and it was an error of judgment, for which they have a measure of responsibility similar, but less, than that resting upon the shoulders of the Division Commanders. There is another aspect of the case. The traditions of the sea are strong, the ideals high, and the rules which seafaring men set for themselves are rigid and hard.

Only by living up to the most rigid standards may the lives of women and children entrusted to the care of seafaring men be safeguarded as far as human effort may make them safe. If a captain loses his ship, he loses his command even when attending circumstances point almost entirely to his complete exoneration from blame. The Navy can do no less. Each captain who loses his ship must bear a responsibility due to that loss. Even though a court honorably acquits him of blame, he must first assume the responsibility for the ship he commanded. Only by maintaining this standard can the high ideals and traditions of the Navy be preserved.

The Old Stern and Spartan Code

The Opinions and Conclusions of the Court of Inquiry were presented in a most masterly fashion. They were sound in every detail. Their setting forth of the responsibility which every commander of a ship or larger unit must, under the Navy Regulations, assume for the ship or ships and the men under his command, is a restatement of something which every naval officer understands and keeps ever in his heart and mind. Command is an honor.

To earn that honor and to wear it in a manner worthy of the traditions of the sea and of the United States Navy is the hope—the dedication—of every man who holds a commission in that loyal Service—and to many who have yet to win their commissions. The area in which the so-called "follow-the-leader" doctrine comes into conflict with this paramount duty—in matters related to proper navigation of his command—is one which must be delineated by common sense.

To that should be added careful study of "Rocks and Shoals," as the Articles for the Government of the Navy were known in those days.

It harked back to the days of wooden ships and iron men and laid down a stern and spartan code in which " ... the punishment of death or such other punishment as court martial may adjudge may be inflicted on any person who ..." committed various crimes. The list included displaying cowardice, "pusillanimously crying for quarter," keeping out of danger to which he should expose himself, failing to engage the enemy, and so on.

Not the least of these crimes was willfully hazarding his ship or suffering her to be run upon rocks or shoals, whereby the lives of the crew were exposed to danger. The penalty for stranding through negligence was, of course, not so severe, the emphasis being on the intent or motivation of the act, which is as it should be. In war, toward which all peacetime training is pointed, the destruction of the enemy is the vital mission. Many risks—some calculated, some on the spur of the moment—are taken which, in the cold light of rules and regulations, might be considered unwarranted. In peacetime, many a wise commander of a force has overlooked and forgiven a miscalculation of considerable importance on the grounds that "you can't make an omelette without breaking a few eggshells."

In war, many a man has been decorated when, actually, there
was a question in his commander's mind as to whether he should be given a medal or a court martial. World War II had many such cases, and the Honda Court of Inquiry in its

conclusions cited two more classical examples dating back to the age of sail, which involved Admiral Lord Nelson of the British Navy.

In the battle of Cape St. Vincent, fought against the Spaniards in 1797, Admiral Jervis commanded a British squadron of 15 men o' war, opposed by 18 Spanish ships of the line. Both fleets were arrayed in the line of battle and Nelson's Captain was the thirteenth in line.

In order to bring his vessels, all heading southward (the Spanish were heading northeast-ward), onto a parallel course and engage the enemy, Jervis either had to make his ships tack in succession from the head of the line which meant making a long and time-consuming U-turn, or he could have his ships wear (go-about) in succession from the rear.

Unfortunately Jervis elected to tack in succession, a slow and time-wasting process. From his position, nearer to the Spaniard's than Jervis, Nelson could see the Flagship Culloden turning and coming up toward the Captain and the Spaniards.

The former was to the leeward of the enemy who was standing north-northeast on a west-southwest wind. Instead of following the other British ships down the long, long leg to the turning point to change his course, Nelson took the Captain out of the line and made for the Spanish van. The change in the tactical situation caused by Nelson's act threw the Spaniards off balance and was primarily responsible for the victory scored by Jervis over a superior enemy force.

In acting against tradition, Nelson took an enormous chance. The line-of-battle concept was so firmly established in the Royal Navy that a former Commander-in-Chief had been cashiered from the service because he failed to preserve the line. What the Court of Inquiry evidently meant, in referring to this incident, was that there are times when combat sailors must defy naval taboos, be they the traditions of the line of battle or the destroyer doctrine's follow-the-leader concept. At Copenhagen, in 1801, Nelson had engaged the Danes, who were at anchor in their own harbor. They stood him off with ferocious fire. After two hours of carnage, Admiral Hyde Parker, who could not get into the fight, lost heart and hoisted signal to break off the action. Nelson was amazed and angry. He felt victory was near. But how could he disobey an order? Then Nelson had an inspiration. He adjusted his telescope, put it up to his blind eye, and trained it on the Flagship.

"Really," he remarked, with mock facetiousness, "I do not see the signal!"

The battle went on and half an hour later the Danish resistance came to an end. Thus, by disdaining tradition at Cape St. Vincent and by disregarding orders at Copenhagen, Admiral Nelson won two battles for England. With all due respect to the parallels drawn by the Court of Inquiry, one cannot help but wonder what the Hero of Trafalgar would have done at 2100 hours on 8 September, 1923, if Horatio Nelson had been a tin-can Captain in Destroyer Squadron Eleven making an engineering run from San Francisco to San Diego.

Meanwhile, on September 23 when the Court of Inquiry was in recess, Memorial Services for the dead of the Delphy and the Young had been held in the huge quadrangle of the Naval Air Station on North Island. Some 10,000 Navy men and civilians attended this impressive tribute to the 23 heroes.

Admiral Robert E. Coontz, Commander-in-Chief of the United States Fleet, presided and expressed the feelings of every man in Navy uniform when he said: "In keeping with the highest ideals of the Naval Service of the United States, our Destroyer crews at Honda upheld the best traditions of the Navy.

"Trained to be ready for the last great moment, those who took our Destroyers to sea, and who died with their ships, were true in every way to the trust we had in them—that they would nobly add to the Navy's heroic tradition.

"At Annapolis, the young men training to be officers, find upon the walls of that great Naval school the pictures, busts and records of enlisted men who have made the Navy proud; and it is these examples, which help fill the young officers with the true spirit which breathes in every Naval man.

It nerves them to duty in the face of death, and sustains them in times of woe. "In our distress for our dead, we are sad; in our admiration for their conduct, we are exalted."

As the Admiral and others spoke, Navy planes droned overhead. But, perchance, the most colorful and significant portion of the ceremonies was lost to the public at large.

Shortly before noon, Destroyer Division 32, led by Commander Walter G. Roper, ComDesDiv 32, aboard his Flagship the

Kennedy stood out to sea followed by the Thompson, the Stoddert, and the Paul Hamilton. Aboard the vessels were the survivors of the Young and the Delphy and on the fantails of the destroyers were great heaps of flowers, literally mountains of them, representing every shade in the rainbow.

When the Division cleared the entrance buoys off Point Loma, course was changed to 170 degrees, true, heading for the dumpling-like Coronados Islands which break the sea horizon to southward. At 1234 hours, with North Island towers in sight and all ships steaming at 15 knots, the Kennedy's rudder was put 20 degrees right and the other three, following the leader, swept across the whitecapped sea in a tight circle only 400 yards in diameter.

Then, on signal from the flagship, the surviving members of the Delphy and the Young began casting flowers over the stern to form a wreath of color on the tossing seas. One might have expected that, at such a time, the ships would reduce speed to emphasize the solemnity of the moment, but not the Cavalry of the Sea. Their shipmates had lived in racing four-stackers, they had died in crashing surf. It was fitting that the gay flowers given to the waters in their memory should toss in foaming wakes. Not until the circle was completed did the speed change. Then, on signal, all engines stopped, all colors came down to half mast, all hands on deck and at stations below stood silent for 2 minutes while the mournful notes of 'Taps" sounded from the bridge of the Kennedy.

"May God rest their gallant souls," was the prayer in the hearts and thoughts of all who heard those golden notes float out over the wide ocean.

Letters of Commendation

As expressed in its opinion, the Court of Inquiry did not find extenuating circumstances to soften its indictments against 11 of the 13 officers who appeared before it as defendants.

But it did recognize that great courage and splendid discipline had been displayed by all hands. However, when it came to making recommendations for letters of commendation, the only Division Commander mentioned was doughty old "Uncle Walt"

Roper, ComDesDiv 31, who had been so what many believed to be painfully outspoken at the trial as to what he would have done and not done if he had had squadron command on the night of the disaster.

The only commanding officers selected for mention were Commander William L. Calhoun of the Young and Lieutenant Commander Walter D. Seed. A longer list of citations, headed by the ComDesDiv 33 and 31, was issued by Rear Admiral Sumner E. W. Kittelle, Commander Destroyer Squadrons, Battle Fleet, Pacific. It failed to mention Captain Edward Watson. Nor did it refer to Commander Roper, who was singled out by the Court of Inquiry. On the other hand, Captain Hunter and Lieutenant (jg) Blodgett of the Delphy—both severely indicted by the Court of Inquiry, were cited for coolness and efficiency during crisis. Similarly, other skippers of wrecked destroyers were given favorable comment for their acts during the emergencies that arose on the heels of the strandings. The list also included about half a hundred enlisted men. Most of the heroic deeds these officers and men performed have already been described. However, they are highlighted here for purposes of well-deserved identification.

"The grief which the Commander of the Destroyer Squadrons feels at the loss of so many brave men, and the further serious loss of seven vessels," the Admiral observed in issuing the citations, "has been in some degree tempered by pride and gratification at the splendid behavior of all officers and men attached to the stranded vessels at the time of the disaster.

"All reports are unanimous in recording the high order of discipline maintained from the moment of grounding until all of the personnel were returned to the base. The attention of the commander of the destroyer squadrons has been invited to numerous special acts of highly meritorious service."

Admiral Kittelle then cited the individual instances.

13 - Facing the End of the Road

Indictments Spread Depression

THE FINDINGS OF THE COURT OF INQUIRY WERE greeted with general satisfaction by press, public and—not least—by Congressional politicians. The approval uttered by Secretary Denby was reflected widely by the economy-minded Coolidge administration.

It was apparent that to certain factions in Washington, the real tragedy of Honda—the loss of seven splendid fighting ships and 23 irreplaceable lives—was of secondary importance. To them the matter for grave concern was that $13,000,000 worth of Government property had been tossed on the junk heap. For that, heads must roll in the sand.

Secretary of the Navy Edwin Denby, of Detroit, Michigan, was on the spot.

He was a serious-minded gentleman of the old school. He was held in high esteem by all hands in the Navy, who rightly refused to believe him blameworthy for the Teapot Dome oil machinations then under investigation in the Senate. He was a fighting man who had served as a gunner's mate at the Battle of Santiago in the Spanish War. (In World War II, he was destined to lose his son, Edwin Denby, Jr., in the submarine Shark, in the Celebes Sea by enemy action.) The Secretary undoubtedly felt that the Navy must lean over backwards, if necessary, in dealing out even-handed justice in the Honda cases.

In his opinion, the honor and the prestige of the Navy were at stake. Any judgments rendered which smacked of undue leniency or a whitewash would do irreparable harm to the Service by bringing on a storm of damning criticism from the public and from the Congress. With the unsavory odor of the oil scandal, which involved illegal leases of Navy petroleum reserves, still in the nostrils of the nation, perhaps Secretary Denby wanted to show that the legal machinery of the Navy was as clean as the proverbial hound's tooth. As for the Navy itself:

The mass indictments handed down by the Court spread a pall of depression over every ship and naval station in the nation. But it was, of course, especially thick and dark among destroyer men and within the confines of the Eleventh Naval District. The broadside fired by Rear Admiral Pratt and his judicial gunners against the 11 defendants inflicted hulling shots between wind and water and left them vulnerable to griping anxieties and black despair.

What would a court martial find? This situation was further aggravated by the sorrow, soul-searching, and misery that engulf any ship commander who, culpable or not, suffers the loss of his vessel.

Could this, for them, be the end of the road?

Leaving for the moment, Captain Watson, Lieutenant Commander Hunter, and Lieutenant (jg) Blodgett out of the picture, the feelings of the Commanding Officers of the destroyers were that the conflicts in concepts, between Navy Regulations and destroyer doctrine, left them no place to jump between the uncompromising frying pan of the former and the disciplinary demands of the latter. If they heeded one set of directives, they ran serious risks of violating another. As one destroyer skipper expressed it: "We are damned if we do. We are equally damned if we don't. We obey orders and follow the leader and, presto, the leader turns into a JAG office tiger and we have it by the tail."

Also disturbing to the morale of all the officers concerned was that they automatically had been relieved of their commands. In order to fill the holes in Squadron 11, brought about by the pile-up, seven deactivated destroyers were restored to active status.

The complements of the wrecked vessels were transferred to the new ships; a new ComDesRon 11, new ComDesDiv 33 and 31 and seven new Commanding Officers were assigned. For a Navy man, a commanding officer of a combatant ship or unit, this was punishment indeed; and it smacked of prejudgment.

With the Navy Department and the press of the nation crying for their blood, what chance had they of acquittal? It must be admitted that human error and poor judgment at the top were the contributing factors in the pile-up. On the other hand, it seems, on the surface, difficult to understand why Commanding Officers who acted under higher orders issued under accepted standard of procedure should be held equally accountable.

During the weeks before the General Court Martial convened, there was considerable speculation on just how far the members of the General Court would go in siding with the concept of total responsibility voiced by the Court of Inquiry.

With respect to the three so-called top defendants, Captain Watson as Squadron Commander and Lieutenant Commander Hunter and Lieutenant (jg) Blodgett, who did the navigating and set the course for the entire column of destroyers, the general opinion was that they did not have very much to hope for in the matter of acquittal.

The Court of Inquiry had brushed aside all testimony that reflected on the efficiency of the radio compass installation. In pointing to the basic factor in the mishap, the Court held that the disaster was "directly attributable to bad errors of judgment and faulty navigation. The price of good navigation is constant vigilance." One may wonder why it was that so many experienced and highly professional sailors not only could be caught in a predicament such as the Tragedy at Honda, but also why not a single soul on the bridges or in the chartrooms of so many destroyers seemed to have the slightest inkling that trouble was brewing.

At the risk of oversimplification, the answer is that the men in command of the destroyers in the Delphy's wake took it for granted that the position signal flashed to ComDes Battle Fleet at 2000 was based upon more definite fixes and better navigational conclusions than was actually the case. Taking for granted that the 2000 position was accurate, there was nothing in the picture to give anyone cause for worry. Its acceptance was an accepted destroyer practice. It should also be borne in mind that, until the Delphy led the way into Honda's coastal fog-curtain, there was nothing to indicate the approach of zero-zero visibility. Running through fog off that part of California was a very common experience. And close shaves were not infrequent.

In fact, only 4 days after the disaster was staged at Honda, the battleship Texas narrowly avoided a serious collision with the northbound Steel Seafarer.

The incident took place some 20 miles west of Honda. The Texas was steaming at 12 knots followed by the battleship Oklahoma. On sighting the merchant vessel through the fog, the Texas altered course to starboard with full right rudder, sounded three blasts on her whistle, backed both engines full speed, and sounded collision quarters. In backing, the Oklahoma and the Texas almost smashed into each other.

A crash with the merchantman was avoided on the swift realization by the ODD that the Texas' bow would clear the oncoming freighter. He signaled full speed ahead on the starboard engine. Even so, the Steel Seafarer and the battleship passed so close together, port to port, that they scraped abaft of midships but incurred only minor damages. The incident of the Texas is mentioned here simply to show that almost zero-zero visibility does not, of itself, stop the movement of ships at fairly high speeds.

General Court Martial Convenes

General Court Martial proceedings were begun against the defendants early in November. It was a seven-member Court presided over by Vice Admiral H. A. Wiley, commanding the Battleship Divisions of the Battle Fleet.

Other members were: Rear Admiral L. M. Nulton, commanding Battleship Division 3 of the Battle Fleet; Rear Admiral J. V. Chase, commanding the Fleet Base Force; Captain De Witt Blamer, senior member of the Board of Inspection and Survey; Captain E. H. Campbell, commanding Mare Island Navy Yard; Captain W. S. Crosley, commanding the Idaho, and Captain S. E. Moses, Assistant Commandant of the 12th Naval District. Also Lieutenant Commander L. E. Bratton, commanding the Stoddert, who continued as Judge Advocate. The trial of Captain Watson began on November 5 and lasted two days. Captain Craven and Lieutenant Commander Weyler acted as counsel.

The dozen witnesses called gave evidence of practically the same nature, that of outlining the events leading up to the disaster and the pile-ups of the destroyers. In the procession of witnesses were Lieutenant (jg) Blodgett and Lieutenant Commander Hunter.

The former testified that, while he aided in the navigation of the Delphy, he had never been designated as the ship's Navigator.

On the stand, "Dolly" Hunter adhered firmly to his contention that the radio bearings had plainly been in error. He admitted that he had disregarded the bearings and said that

Captain Watson, at all times, had been fully informed on the navigational situation. Summing up for the prosecution, Lieutenant Commander Bratton argued that Captain Watson was responsible for the safe navigation of the vessels in his Squadron.

He held that the fact that his vessels were wrecked indicated that Watson had been inefficient in the performance of his duty. Finally that the Court must decide if the radio bearings received indicated that the Delphy was too far north of Point Arguello to make the turn ordered in safety. In answer, Lieutenant Commander Weyler declared that the evidence showed that the Squadron Commodore had not neglected, as alleged in the specifications, to heed the radio bearings received or the possible necessity for reducing speed and taking soundings.

Commander Weyler, in defining the word "heed," said that the accused had heeded these matters and had given them every consideration, and that his mistake in judgment was made while all these facts were in mind.

While he admitted that there had been a grave error in judgment, he also argued that it was not due to inattention to duty or inefficiency in performing his duty. Captain Craven summarized the arguments for the defense.

"The error of judgment," he said, "was made by men who were, at the time, intent upon doing their full duty."

He contended that the methods followed were those "ordinarily observed by naval men making voyages in destroyers, and that there were unforeseen contributory causes that resulted in the various circumstances that led up to the disaster."

Testimony had been unanimous, he recalled, that the speed and course were safe under the conditions obtaining; that an adverse and unexpected current had been encountered; that the radio bearings had been erratic and difficult to obtain, and that the wreck of the passenger vessel Cuba had caused further congestion of the radio situation and had demanded Captain Watson's attention.

He recited the evidence that the change of course ordered was not of itself considered dangerous. He closed by recounting the unusually fine service record of Captain Watson and drew attention to the Squadron Commander's courageous act in taking full responsibility for the strandings.

The hearing room was cleared so that the Court might meet in executive session. When the Court reconvened it voiced its readiness to hear the case against Lieutenant Commander Hunter. The fact that the Court did not announce its verdict in the Watson case was taken to mean that the Captain had been found guilty. Otherwise, conforming to custom, the Court would then and there have declared the defendant not guilty. At the outset of Lieutenant Commander Hunter's case, Captain Watson, called by the prosecution, became Hunter's star witness. He paid tribute to Hunter's skill as a navigator.

And, while Watson admitted that, if the Delphy had been slowed to 5 knots and soundings had been taken, the flagship and the other vessels would have been saved, he reiterated that there was no apprehension of danger or of the necessity of slowing or taking soundings. The course, if made good, he maintained, would have put the Delphy in water too deep for sounding.

The slowing of the ship after the turn at 2100 could not have saved her, he said, as she crashed too soon afterward. Captain Watson declared emphatically that Commander Hunter "did consider and heed the bearings received" and all other navigational data before recommending the change in course. He added that, if the ship had been where they had every reason to think she was, a change was necessary to avoid San Miguel Island.

After several witnesses for the prosecution, virtually repeating their testimonies before the Court of Inquiry, had given evidence, Lieutenant Commander Weyler, as counsel for the defense, admitted the wrecking of the Delphy, but submitted that he would show that there were certain mitigating circumstances, and that there were contributory agencies that could not be foreseen.

Captain Robert Morris, commander of Division 33; Commander Louis P. Davis, commander of the Woodbury, and Commander William L. Calhoun, commander of the Young, appeared as defense witnesses.

They were unanimous that the speed and the course were both safe, and that normally the turn ordered could have been considered safe. None of them expected any danger.

Captain Morris said that he would have preferred to make soundings, but that he did not request that they be taken.

He also said that he personally would have reduced speed, not because of the danger of striking the shore, but because a large number of merchant vessels could be expected to be found in Santa Barbara Channel.

The defendant concluded his own case by taking the stand and once more going over the navigational testimony he had given during the trial of Captain Watson, which was almost identical.

He assumed full responsibility for the navigation of the Delphy, and announced that he had never detailed Lieutenant Blodgett as Navigator, that he himself did all the navigational work, and that, in the navigation department, Lieutenant Blodgett did only the routine work.

Blodgett Not the Delphy's Navigator

In view of the manner in which Lieutenant Blodgett had been catapulted into the case as the thirteenth defendant by the Court of Inquiry, this admission from Commander Hunter created considerable surprise.

Speaking for the defense, Lieutenant Commander Weyler maintained that the assertion that all of the radio bearings received indicated that the Delphy was too far to the northward of Arguello to change course had not been proved.

He said that there was no evidence to show that Lieutenant Commander Hunter had failed to heed or consider the bearings or any other navigational data. Weyler admitted that the results proved an error, but argued that a quick decision was necessary, that a fuller consideration of the data was not possible at the time, and that Commander Hunter was not guilty of culpable inefficiency. The court made no announcement of its verdict after the trial was concluded. Thus another guilty verdict was indicated.

Next came the trial of Lieutenant Blodgett.

When it opened, Commander Henry M. Jensen, as counsel for Lieutenant Blodgett, announced that the defense would offer no witnesses.

Disdaining Hunter's testimony to the contrary, the Judge Advocate, in final argument, held that the evidence showed that the accused actually was Navigator, and that he had not performed his duty properly.

After a short deliberation, the Court returned a verdict of acquittal. With the three cases involving "culpability" disposed of, the way was clear for the airing of the somewhat lesser charges "of suffering naval vessels to run on the rocks" against the remaining eight defendants.

First to be called was Captain Bobert Morris, ComDesDiv 33. All of the witnesses who had served with Captain Morris on his Flagship and those who had occasion to observe him following the disaster at Point Honda, testified that his conduct had been exemplary, and in accord with the highest traditions of the service.

Taking the stand in his own behalf, the defendant, among other things, stated that he personally blew warning whistles when the ship struck, and that he did everything in his power to save the vessels of his division when the danger became apparent.

The record of Captain Morris was read into the record by Lieutenant Commander Bratton. It contained many highly complimentary statements regarding his ability and devotion to duty, and letters of commendation signed by Admiral Gleaves, Admiral Sims and Secretary of War Newton D. Baker.

When the trial came to an end, the Court announced its finding of not guilty, and the full and honorable acquittal of the Captain. The Court included in the acquittal a paragraph which specified that the verdict was not to be considered, in any way, a precedent which might lessen the responsibility of a Division Commander for the safety of the vessels under his command.

The case, as had been pointed out by both the defense and the prosecution, was without precedent in the annals of Navy jurisprudence.

It involved issues of policy, tactical doctrine, and the allocation of responsibility which were so tangled that it was predicted the evidence might result in changes of Navy Regulations to make them more specific in the direction of large units of destroyers. Both the prosecution and defense, and several of the important witnesses, were agreed that amplifications and changes of Navy Regulations were imperative. The not guilty verdict found for Captain Morris served, like a life-giving breeze, to lift the pall of depression that hung over the Navy. It made prosecution-minded politicians in Washington sniff the air with dismay and wonder.

In San Diego, the feeling prevailed that the case of Captain Morris, being, in a manner of speaking, a bellwether for officers in his own defense bracket, indicated the Court's attitude toward the charge of negligence under the circumstances that prevailed on September 8. The brevity of the trial of Commander Calhoun, commander of the Young, who was next in line, tended to give strength to the rising spirit of optimism.

After several witnesses had reiterated previously reported testimony, Commander Calhoun took the stand. He stated that he himself had navigated as carefully as though he were not in the least dependent upon the ships ahead of him, and stated that he had every reason to believe that the Delphy was doing exactly as he would have done.

In fact, he said he did everything that he would have done had he been alone except for the taking of soundings and the obtaining of radio bearings, both of which navigational aids were forbidden to him because of battle fleet and destroyer squadron regulations and instructions.

A brief executive session by the Court. Defendant not guilty. Next case.

It proved to be that of Commander Louis P. Davis, commanding the Woodbury. Again a procession of witnesses speaking their lines.

An interesting moment in the dullness of the proceedings came when Commander Davis, while on the witness stand, was asked to define his conception of destroyer doctrine. He answered that, in his belief, "destroyer doctrine is the unwritten understanding that develops between the officers of the Force and which enables them to keep constantly of one mind on the matters of tactics and policy."

Commander Davis stated that his own navigational policy was to endeavor to keep the position of his vessel constantly fixed, to do as much navigating as is done on the largest ships, and even more if possible.

He added that he followed this policy whether he was operating alone or in the company of other vessels. If he had been proceeding singly, he said, he would have slowed and taken soundings.

As it was, he obeyed orders, confident that the Delphy would not have continued on the course at the high speed of 20 knots if her officers had not known exactly where they were.

Following summing up by the prosecution and the defense, the Court held a short executive session. At its conclusion, the President announced that the finding in the case of Commander Davis was not guilty. Next to stand trial was Lieutenant Commander Herbert O. Roesch of the Nicholas.

His case opened when the destroyer's Chief Radioman T. C. Carnahan and Radioman F. F. Brown testified that three bearings were intercepted.

Both operators said that the receiving set aboard the Nicholas was not up-to-date; that the guarding of two waves was difficult because of the large amount of interference the set permitted, and that the interception of bearings on a third wave was even more difficult. The Nicholas was further handicapped by having only two operators, who were forced to stand "watch and watch," 4 hours on and 4 off. Neither operator was able to intercept the Delphy's 8 o'clock position, they testified.

Lieutenant (jg) H. F. Sasse, Executive and Navigator of the Nicholas, went through the long process of laying out the courses on the charts, plotting bearings and positions. He stated that Commander Roesch checked his navigational work, that he was on the bridge frequently during the day, and that he even stood one watch in three. Lieutenant Sasse confirmed the operators in their statements regarding the radio equipment. He stated that the navigational procedure of the Nicholas was always the same, regardless of whether the vessel was operating singly or in company with the squadron.

The Court, after hearing additional witnesses and the final arguments of the prosecution and defense, deliberated for 1½ hours, but did not announce its verdict upon reconvening.

Unexpected Upset in Roesch Case

Unfortunate "Fats" Roesch. His was the seventh case before the Court. But on that November 20, seven was not his lucky number. The General Court Martial had found him guilty.

Just why the members of the Court felt that there was a difference in conduct between Lieutenant Commander Roesch and his brother skippers sufficient to justify a guilty finding against him is hard to understand.

Since members of a Court Martial are not allowed to reveal their vote—except in a legal action—those for and against his acquittal will probably never be known. It is noted that, in his official report, Captain Roesch stated that visibility was poor. In the testimony of his radiomen it was stated that three radio bearings had been intercepted—which was more than other ships reported. Possibly the Court felt that, in visibility which he considered poor, he should have made more use of his radio bearings—disturbing as they no doubt were—and perhaps have requested permission to take soundings.

Such a request might have started a chain reaction of questioning which might have produced results. If such were the reasoning, it was obviously unfair to Roesch, and the Commander-in-Chief, Battle Fleet, Admiral S. S. Robison, promptly disapproved the conviction.

When asked years later, why he had been convicted, Captain Roesch replied, "I believe my conviction was because I stated that I was following the leader as, of course, was everyone else. I kept track of the navigation as well as possible but we had nothing to go on but dead reckoning. We couldn't take soundings and were forbidden to ask for radio bearings."

"Anyhow," he added, "when my conviction was disapproved and all of the acquittals were disapproved, we were all in the same boat.

The principal effect of these reversals was that all the skippers were put in the position of having been responsible for the loss of their ships. This prevented us from submitting claims for the loss of our personal gear and equipment as did our junior officers." Thus did Captain "Fats" Roesch, of the sunny disposition and ready smile, brush aside thoughts of what effect the disaster might have had on his career. When the case of Commander William H. Toaz, Captain of the S. P. Lee, Flagship of ComDesDiv 33, was called, much of the evidence given in the trial of Captain Robert Morris came once more to the fore.

Navigational procedure aboard the destroyer was discussed at length, as were conditions of visibility and speed.

The Court's verdict was not guilty!

In the trial that followed, of the case against Commander William S. Pye, ComDesDiv 31, the prosecution ran into shoal waters. It turned out that, with only one exception, the witnesses called by the government gave testimony valuable to the defense under cross-examination. Lieutenant Commander J. F. McClain, skipper of Pye's flagship, the Farragut, said that the Destroyer Division Commander did everything that could be done to save the ships.

He was followed by Commander Walter D. Seed of the Fuller and Commander Richard H. Booth of the Chauncey. They declared that, in their opinions, Commander Pye had done nothing that in any way contributed to the loss of their vessels, and that he had not omitted to do anything that might have saved them. Lieutenant (jg) L. L. Hunter, Acting Division Communications Officer, stated that Commander Pye was at all times intensely interested in the navigation of the ship and spent most of his time on the bridge, while his conduct at the time of the disaster was above reproach.

Captain James H. Tomb, Commander of Destroyer Squadron 12, declared that, in his opinion, the fleet regulations regarding the handling of divisions were developed largely for units of larger ships with more and better trained personnel; that they were valuable for the guidance of destroyer divisions, but that they could not be followed too literally.

He stated that a Division Commander, having no special navigational aids at his disposal, must depend on the faculties of his flagship and its personnel, and that as far as obtaining radio bearings was concerned, a Division Flagship had no preference over any other vessel of the Squadron, but must follow the same rules and suffer the same limitations.

Captain Watson testified that he had always held the Division Commanders responsible for their units, expecting them to keep informed at all times as to the positions of their ships and holding them responsible for their safety. He admitted, on cross-examination, that Commander Pye had never failed to use proper initiative, that he had handled his division well in tactical maneuvers, and that he had already commended him for his conduct on the night of the disaster.

Commander Pye, on taking the stand, said he kept in constant touch with the navigational situation all the way. He declared that he had no warning of any kind that there was danger ahead until the confusion ahead became evident.

"The only thing that was apparent," he said, "was that there had been a collision. I never saw any red lights and heard no whistle signals. All I saw was the Young lying on her side and the lights of two ships ahead. I heard the cries for help that came from men in the water and clinging to the oily sides of the Young. My concern then was, naturally, to save these men, for there was nothing to warn me that my ship was in danger, too.

"I believe, gentlemen," he said after a reflective pause, "that I did everything I could do under the circumstances. I signaled the other ships and hailed them as soon as the danger became apparent to me. If the Somers, Chauncey, and Fuller had received from the Delphy the signal 'Ground ahead' at the time they received the signal 'Nine turn' I believe the whole of the Thirty-first Division would have been saved."

Counsel for Commander Pye offered lengthy arguments in summing up the evidence before the Court. Lieutenant Commander Bratton closed with a brief reply and a summary of the prosecution's case.

A verdict of acquittal was reached by the Court in short order. Lieutenant Davis, Executive and Navigator of the Fuller, was called as witness for the prosecution in the case of Commander W. Dudley Seed, Captain of the Fuller.

He gave the navigational information that was given in each of the preceding trials. He admitted that no soundings had been taken because it was not practicable to take soundings at the speed of 20 knots set by the Squadron Commander. To have reduced speed and to have left the formation for the purpose of taking soundings would have been disobedience to orders, he maintained.

On being recalled as a defense witness, Lieutenant Davis stated that his commanding officer was "right on the job" at all times, and that he was constantly on the bridge at sea. Taking the stand in his own behalf, Commander Seed told the story of the navigation of his vessel from leaving San Francisco until she struck. He stated that he would have liked to have radio bearings. Yet he made no attempt to get them because the instructions were so explicit in forbidding any but the senior ship, or one designated by the senior, to signal for bearings. He observed that it would have been a violation of radio instructions to unguard one wave and intercept bearings. Furthermore, he stated that he had every reason to believe that the Delphy was obtaining bearings or she would not have maintained her high speed and that of the other ships of the Squadron under the poor visibility conditions then obtaining.

Verdict—not guilty.

The trial of Lieutenant Commander Richard H. Booth, on a charge of negligence in permitting his vessel, the Chauncey, to run on the rocks was the last in line. Lieutenant C. V. Lee, Executive and Navigator of the Chauncey, outlined the navigational procedure employed. He told of receiving the signal for a "Nine turn" just before confusion was noted ahead.

"This turn was partly executed," he said, "when rocks were seen ahead.

This was the first indication of danger, as it was thought that the confusion was due to a collision and that the 'Nine' turn had been ordered to prevent following ships from being involved in the collision.

By the time the rocks were noticed, it was impossible to extricate the Chauncey.

"If the signal 'Ground ahead' had been received instead of 'Nine turn,' the Chauncey would have backed out of trouble and would have been saved."

Lieutenant (jg) Lawrence Blodgett and Lieutenant Commander R. E. Bell testified regarding radio bearings received by the Delphy and the Kennedy, respectively. Further radio information was obtained from Radiomen F. W. Fish and W. E. Mann of the Chauncey. Both testified that they had been unable to intercept radio bearings while guarding the prescribed waves, though Commander Booth had frequently inquired of them if any bearings had been or could be intercepted. Commander Booth then took the stand himself. He testified that the signal "Ground ahead" would have saved his vessel, but that it was too late to do anything after the first warning and the rocks were sighted and struck.

Commander Booth was the last defendant before the tribunal. He received a not guilty verdict.

Not Guilty Verdicts Disapproved

The seven-member General Court Martial was dissolved after having handled the largest number of cases ever considered by any single General Court Martial in the history of the Navy. The results were three convictions and eight not guilty verdicts. As was later announced, Captain Watson was sentenced to the loss of 150 numbers on the list of Captains.

Lieutenant Commander Hunter was sentenced to the loss of 100 numbers on the list of Lieutenant Commanders. This meant that neither of them would ever reach eligibility for promotion. They were not, however, deprived of the right ever to command a naval vessel again. This bar, effective in the Merchant Marine in similar cases, does not exist in the Navy.

Ordinarily, this would have been the end of the matter except for the standard enshrouding procedures with JAG office red tape. But the Tragedy at Honda was not to leave the boards that easily. While the findings dissipated the gloom that hung over the Navy, they created deep displeasure in Washington. The wholesale "hangings" that had been expected did not materialize to provide material for political campaigners.

There was much talk about drastic Congressional changes in laws that deal with naval courts, to place greater legal power in the hands of the Secretary of the Navy. Even Calvin Coolidge—who seldom ventured an opinion on any subject, including the weather—observed that the "Court Martial has been very lenient with everybody."

As is understandable, Secretary of the Navy Denby was much displeased with the eight acquittals. He believed that stern treatment should have been accorded all 11 of the defendants. He felt that the standards of naval discipline had been let down and that the prestige, performance, and morale of the Service would suffer. However, there are other angles to be considered. It was obvious that the Division Commanders and Commanding Officers involved had placed too blind a confidence in their Squadron Commander.

Also that he had exercised too dictatorial a control over their freedom to navigate the ships for which they were responsible. Sober reflection upon and consideration of all elements involved had been given by the members of the Court Martial and the verdicts arrived at came from the depths of their collective conscience. A newspaper man once made the remark during the course of the nationally publicized trial of an Army officer, "If I were guilty, there is no court I would rather not be tried by than a military court martial; if I were innocent, there is no other court which I would prefer."

Alongside that remark should be placed the old axiom of naval courts: "We deal with justice, not law."

Placing absolute faith and confidence in a leader has been both a curse and a blessing since the dawn of history. It has led to the rise of despots and dictatorships; it has led to the creation of free nations, such as ours; it is the foundation of all religions. Again the dividing line must be drawn by common sense, conscience, and evaluation of the end to be attained. No leader or ambition should ever be followed blindly unless the risk has been carefully calculated.

In war calculated risks are matters for everyday consideration. Before the records of the Court Martial came to the desk of Secretary Denby they passed through the office of the Judge Advocate General of the Navy, where that officer prepared an endorsement recommending disapproval of the acquittals of Captain Morris and Commander Pye, ComDesDiv 33 and 31, as well as destroyer commanders Toaz, Calhoun, Davis, Seed, and Booth.

The overriding by Admiral Robison of the verdict against Commander Roesch was likewise blasted. Early in 1924, the not guilty verdicts were disapproved by endorsements placed on the Court Martial records by the Secretary of the Navy. This action was made public by the Navy Department with a statement noting that disapproval of the findings by the Secretary did not serve as a basis for re-trial of the cases, but was "simply an expression of the Secretary's views of the Court's action."

In the cases of the two Division Commanders who were acquitted "fully" of the charges, the Secretary adopted the recommendation of disapproval made by the Judge Advocate General in the indorsement in which he held that the "evidence adduced not only fails to warrant the action of the Court in 'fully' acquitting the accused, but also fails to warrant an acquittal."

Similar action was taken in the case of the destroyer commanders who were acquitted on the negligence charge.

The Judge Advocate General held that the evidence in each case "shows clearly that although the accused had grounds for believing that the position of the vessel under his command was in doubt and that the course directed by the Squadron Commander was leading the ships into danger, he did not give notice of that fact to the officer in direct authority over him or to the other ships endangered."

What Did the Future Hold?

The effect of this action was to leave the records of the officers concerned under a legalistic cloud. To analyze whatever effect, if any, it had on their careers would be the wildest sort of guesswork. Perhaps the best way to present the post-mortems of the trials is by way of a simple review of the records of those who were involved during the years that followed the Honda disaster.

Captain Edward H. Watson, born in Frankfort, Kentucky, in 1874, who suffered a loss of 150 numbers on the list of Captains, served as Assistant Commandant of the Fourteenth Naval District at Pearl Harbor, Hawaii, until 1929 when he retired. Oddly enough, the vacancy left by Captain Watson's retirement permitted the promotion of Commander William L. Calhoun to the rank of four-striper. Captain Watson died in 1942.

Lieutenant Commander Donald T. Hunter, born in Elgin, Illinois, in 1887, who suffered a loss of 100 numbers on the list of Lieutenant Commanders, served as Navigator of the battleship Nevada, as First Lieutenant of the Oklahoma, and as instructor at the Naval War College. He, too, retired in 1929 and died in 1948. Captain Robert Morris—born in Atlantic, Iowa, in 1878—served 2 years as commander of the Navy transport Henderson. He then commanded the Rigel and finally for 2 years he was the Captain of the battleship Maryland. He held other important assignments until he retired in 1935. He died in 1955. Commander William H. Toaz—born in Rochester, New York, in 1883—was first assigned as Executive Officer of the hospital ship Relief. Other assignments included a year at the Naval War College and command of the Cuyama of the Training Squadron of the Fleet Base Force. He retired in 1930 and died in 1944.

Commander William L. Calhoun—born in Palatka, Florida, in 1884—served as Navigator of the battleship Maryland. Later assignments included ComDesDiv 31. Over the years, Admiral Bill Calhoun built up an impressive record and a distinguished career. He holds many commendations, but the one of which he is proudest came to him in December 1923, from Admiral Lewis Bayly, R.N., who commanded the American destroyers that were based at Queenstown during World War I. In commenting on the part Commander Calhoun played in the Tragedy at Honda, this famous sea-dog among British Admirals wrote:

"It is a story of calm courage, perfect discipline, and faith in God, and has done me a very great deal of good to read it. I like so much the order to be ready to abandon ship. Some would have abandoned her and lost their lives! I hope an Englishman would have given the order you did: an American did it. The discipline must have been as near perfect as could be; everyone trusting in you and waiting for your order, and then to thank God publicly for his rescue. Thank you for your splendid example. Yours sincerely, Lewis Bayly. Very well done!"

Vice Admiral Calhoun was charged with the logistic support of the Pacific Fleet and naval shore-based establishments in the entire Pacific area during World War II. He was retired in 1946 after 44 years of active Navy service with the rank of four-star Admiral. At this writing, he lives in Coronado, California. One of his proudest possessions is a sword presented to him by the survivors of the Young. This sword, after his retirement, he presented to a midshipman just graduating from the Naval Academy in order that it might continue to serve in the Navy.

Captain Louis P. Davis—born in Wilmington, North Carolina, in 1883—served as Executive Officer of the battleship Colorado, Squadron Commander of DesRon One, and Captain of the Maryland. His assignments also included a tour of duty at the Naval War College. He retired in 1940 and currently lives in his home town, Wilmington, North Carolina.

Captain Herbert O. Roesch—born in Pendleton, Oregon, in 1886—served, among other berths, as First Lieutenant of the battleships West Virginia and California. A nationally famous rifle shot, he did a 2-year tour as Captain of the Navy Rifle Team at Annapolis. During World War II he was a Convoy Commodore on the Western Sea and Hawaiian Sea Frontiers. He retired in 1945 and now resides in Coronado, California.

Commander William S. Pye, born in Minneapolis, Minnesota, in 1880, served in the War Plans as well as the Naval Intelligence Division of the Office of the Chief of Naval Operations. He was known as an authority on strategy and planning for war. In 1936, he became Assistant Chief of Naval Operations with the rank of Rear Admiral. In 1938, he lost his son, William S. Pye, Jr., in the crash of a Navy fighter plane. Following other assignments, he became Commander Battleships, Battle Force, with rank of Vice Admiral and, briefly, after the attack on Pearl Harbor, served as Acting Commander-in-Chief of U.S. Pacific Fleet pending the arrival of Admiral Nimitz to take over.

Vice Admiral Pye's active service ended in 1945 at which time he was President of the Naval War College. Admiral Pye died in 1959. Lieutenant Commander William D. Seed—born at Tuscaloosa, Alabama, in 1888—served as Exec, of the Gold Star. After a few months of service he was hospitalized and transferred, in August 1924, to the Retired List for reasons of ill health. He died in San Antonio, Texas, in 1957.

Commander Richard H. Booth—born in Harlan, Iowa, in 1889—among other duties, served as instructor at the U.S. Naval Academy and the U.S. Naval War College. His seagoing assignments included the Wyoming, the John D. Edwards (commanding), and the Black Hawk (Exec). Commander Booth retired in 1938 after serving two years in Headquarters, ComTwelve. He now lives in Oroville, California.

Lieutenant (jg) Lawrence Blodgett—born in Stoneham, Massachusetts —won his Ensign's commission in 1918 after active World War I duty. He was promoted to a full lieutenancy in 1926 and placed on the retired list, with rank of Lieutenant Commander, in 1941. He continued on active duty until 1947. His commands included the Avocet, the Mahopac, and the Whippoorwill. During and after World War II, he commanded Naval Ammunitions Depots in Hawaii. Lieutenant Commander Blodgett resided in Honolulu, where he died in 1958.

Many of the junior officers who showed such promise of leadership and cool courage when their destroyers ran on the rocks at Honda, went on to serve with great distinction in the Navy. Among those who made flag rank were: Lieutenant (jg) Richard H. Cruzen of the Delphy—Vice Admiral; Lieutenant (jg) Allen P. Mullinix, also of the Delphy—Rear Admiral; Lieutenant (jg) Paul W. Steinhagen and Ensign William D. Wright, Jr., of the S. P. Lee— Rear Admirals; Lieutenant (jg) Felix L. Baker of the Young—Rear Admiral; Ensign William W. Juvenal and Ensign Horatio Ridout of the Woodbury—Rear Admirals; Ensign Leo L. Page of the Chauncey; Rear Admiral Lieutenant Commander Leslie E. Bratton, skipper of the Stoddert, who served as Judge Advocate at the trials, also became a Rear Admiral.

Lieutenant Commander C. E. "Granny" Cobb, who did such a fine job soft-shoeing the Percival out of harm from Honda's rocks, made Vice Admiral. Rear Admiral Bratton is a resident of Denver. Vice Admiral Cobb lives in Chevy Chase, Maryland. Ensign C. C. Hartman of the Farragut is currently Commandant of the Eleventh Naval District with rank of Rear Admiral.

Nearly four decades have passed since that September day when Destroyer Squadron 11 steamed down along the California coast at 20 knots, hellbent for Honda. Since then, the jinns of wind, wave, current, and fog have conspired to lure many more luckless sailors into the Graveyard of Ships—passenger steamers, freighters, and other types of craft. Neither brighter beams nor louder foghorns at Point Arguello Light—nor greatly improved radio bearing devices—have made this dismal stretch of coast safe for the unwary or careless seafarer. Since then, there have been no major changes in the laws that govern naval courts, except that the Universal Code of Military Justice now governs all Armed Services. Neither have the imbalances between Navy Regulations and destroyer doctrines been removed as was suggested during the General Court Martial of the 11 defendants and when Vice Admiral H. A. Wiley, President of the Court, made this terse comment:

"Navy Regulations should say what they mean—and mean what they say!"

As for the jinns, they wait in their lairs, with the placid patience of immensity, for new generations of sailors to venture across the edge of hidden peril into La Guijada del Diablo.

Honda!

THE CHAUNCEY RUN AGROUND

Made in the USA
Las Vegas, NV
12 January 2023

65506628R00163